P9-CRO-608

FOSSILS IN COLOUR

Companion volumes:
MINERALS AND ROCKS IN COLOUR
FIELD GEOLOGY IN COLOUR

FOSSILS
in colour

by J. F. KIRKALDY
Emeritus Professor of Geology
Former Head of the Department of Geology
Queen Mary College, London

photographs by
MICHAEL ALLMAN

BLANDFORD PRESS
POOLE DORSET

First published in the U.K. 1967
Second edition (revised) 1970
Third edition (revised) 1972
Fourth edition (revised) 1975
Reprinted 1980

Link House, West Street,
Poole, Dorset, BH15 1LL

ISBN 0 7137 0743 7

Printed in Great Britain by
Fletcher & Son Ltd, Norwich
Bound by Richard Clay (The Chaucer Press) Ltd
Bungay, Suffolk

CONTENTS

The maps and text illustrations are by Victor Farmer

PREFACE

This book is a companion to the author's *Minerals and Rocks in Colour*. In this case, the fine colour photographs are the work of Mr. M. Allman, formerly Chief Technician to the Geology Department, Queen Mary College and now Lecturer in Applied Scientific Photography at Harrow Technical College. The specimens illustrated have been chosen from the collections at Queen Mary College. A number of the specimens are not perfect, but this is deliberate, for they are intended to illustrate the kind of material that one is normally apt to find rather than the exceptionally perfect specimen which should be housed in a national museum. In the majority of cases, the locality from which the specimen came can be given and unless otherwise stated, it is in England. But in some cases the locality is not known. These specimens were acquired from old collections made perhaps a hundred years ago. At that time certain fossil collectors were deliberately secretive about their best localities or another frustrating alternative is that during the passage of time, and perhaps as a result of many handlings, labels have become detached from the specimens and lost. This underlines certain of the points made in Chapter V.

In the text the author has attempted to treat fossils on a very broad front. To show both their fascination and the practical value of their study, to explain why particular fossils can only be found in certain areas, and to regard them not as interesting or perhaps bizarre shapes, but as the remains of organisms that were once alive. In Chapter IV, only a limited number of genera from each of the main groups of fossil organisms are described and illustrated. The greater number of those chosen are common fossils, which show the range of variation within a group and are also of value in reconstructing the conditions of the geological past. References are given in Chapter VI and in the Bibliography to more comprehensive texts, mainly concerned with the morphology (shape) of fossils, which will enable one to identify more precisely the many fossils, both common and rare, which cannot be mentioned or illustrated here in the space available.

As well as to Mr. Allman, the author would like to acknowledge his debt to his colleague Dr. F. A. Middlemiss for his constructive criticism of the text, to Mr. V. Farmer who has drawn nearly all the illustrations, to Miss M. Knight, who turned his illegible manuscript into faultless typescript and to the publishers for their help and forbearance.

J. F. Kirkaldy
July 1967

Geology Department
Queen Mary College
University of London

This second edition has provided the opportunity to improve the colour section. Mr. Allman has taken a number of new photographs, whilst others have been printed at greater enlargement. A number of minor changes have been made in the text to incorporate the latest work, particularly on the time range of certain fossils, whilst the section on Literature and the List For Further Reading have been substantially amended.

In this third edition further minor revisions have been made in order to keep the material up to date.

It is the convention in palaeontological literature to print the generic and specific names of fossils in italics. So as to facilitate the use of the General Index, the first appearance in the text of the names listed there have also been italicized, whether they be technical terms for the parts of fossils (e.g. **muscle scars**), significant groups of fossils (e.g. **orthids**) or important rock units (e.g. **Wealden Beds**).

I. THE NATURE OF FOSSILS

The carefully arranged fossil sea-urchins from the Chalk found in certain Bronze Age graves show that about ten thousand years ago some men must have regarded these fossils as unusual and as worth collecting, possibly for use as charms. But it is only within the past two centuries that the true nature and value of fossils has been appreciated.

It is true that some of the classical scholars and philosophers, such as Herodotus (484–425 B.C.), recognized that fossils were the remains of once living organisms and that the presence of fossilized marine shells in Malta, Egypt, etc., showed that these areas must have been submerged beneath the sea. But during the Dark Ages very different ideas were put forward. Fossils were regarded as sports of the devil placed in the rocks to mislead mankind; they were due to Noah's flood; they were 'sports of nature'; they were produced by a life force trying unsuccessfully to make living organisms out of rock; they were peculiar mineral formations or formed by seeds carried by the wind from the seas, etc., etc. Even the word fossil (from the Latin *fossilis*, something dug up) as used by its originator, the German physician Agricola (George Bauer 1494–1555), included minerals as well as what we now regard as fossils. Indeed Agricola's treatise *De Natura Fossilium* is really the first textbook of Mineralogy, for in it he described with considerable accuracy the various minerals to be found in Saxony, at that time the chief mining area of Europe.

Since then the term fossil has become limited to the remains and traces of the life of the geological past, though in the sense of 'former' it is sometimes extended to include such inorganic accumulations as fossil sand dunes, fossil beach deposits, a fossil coast line, etc. Whilst the potsherds, coins, buried buildings excavated by archaeologists are strictly speaking fossils, the term is not normally applied to them, for the geological past is regarded as ending at the beginning of the Holocene Period (p. 17), that is about ten thousand years ago. The bones, tools, etc., studied by archaeologists are too recent to be included within the geologist's concept of fossils.

During the seventeenth and eighteenth centuries interest in fossils lagged well behind the steadily growing rise of many branches of science and it was not until the close of the eighteenth century that the value of fossils was shown by the work of William Smith (1769–1838), the 'Father of Stratigraphical Geology'. Up till then it is true that fossils had been collected, illustrated and described, but little attention had been paid to the rock layers in which they occurred. William Smith, an engineer by profession, was engaged on the construction of the Kennet–Avon canal. He studied the rocks exposed in the cuttings, collected their fossils and by keeping separate the fossils found in the different rock layers, was able to show that 'strata can be identified by their organized fossils'. The significance of Smith's dictum will be considered more fully in the next chapter.

During the early years of the nineteenth century the foundations of **Palaeontology**, the study of the life of the past (after the Greek for discourse (*logos*) on ancient (*palaios*) beings (*onta*) were laid by numerous workers, amongst them the French scientists J. Lamarck (1744–1829) and G. Cuvier (1769–1832), working on invertebrates and vertebrates respectively. The Frenchman A. D'Orbigny (1802–1857) and the German A. Oppel (1831–1865) paid particular attention to the chronological value of fossils. The publication in 1859 of Charles Darwin's (1809–1882) *Origin of Species* was another landmark, for most of the evidence which he advanced for evolutionary change came from fossils. His work naturally led to much more detailed studies than hitherto of the changes undergone during the passage of time by numerous groups of fossils.

The scope of palaeontology has steadily widened, particularly during the last twenty years. Not only are new types and varieties of fossils constantly being found but fossils are being studied from new angles. As the search for oil became increasingly guided by scientific methods, more and more attention was paid to the study of the fossils, especially of the minute *microfossils*, obtainable from the cores and cuttings brought up by the oil drill. Millions of pounds per annum are now spent by oil companies in palaeontological work alone. Another change is the increasing attention now being paid to palaeoecology, that is the relation between fossils and the environment in which they lived. To quote an eminent American vertebrate palaeontologist, 'Fossils should be regarded not as things dead and buried, but as forms of life.' Again there is the increasing recognition that palaeontologists and neontologists (the students of present life) should combine in the study of problems such as evolutionary processes, to which each can bring his own distinctive contribution.

Fossilization

It is very exceptional for fossils to be more than the hard parts of organisms that lived in the past – that is the shells of bivalves, sea-urchins, etc., the teeth and bones of vertebrates, the trunks, branches and leaves of trees. The soft parts have disappeared, owing to the rapidly acting processes of bacteriological decay aided often by the attack of scavengers of various kinds. These can only be prevented by the sealing up of the complete organism almost immediately after its death, as has occurred in the remains of the extinct woolly mammoths (*Mammathus* (*Elephas*) *primigenius*), which have been preserved for more than thirty thousand years in the frozen ground of Siberia or, another example, the extinct woolly rhinoceros (*Rhinoceras tichorhinus*) which was found in southern Poland, sealed up in asphalt marking the site of a former tarpool. Some of the mammoths had been preserved so well in cold storage that their flesh was edible, at least by sledge dogs, whilst their stomachs were full of undigested grasses. But such finds are due to a most exceptional combination of circumstances and also have only been made in rocks of recent age, geologically speaking.

No. 1 shows a polished cross-section cut through an ammonite (p. 143). Ammonites secreted a spirally coiled shell divided into a number of chambers separated by septa running from the inner to the outer walls of the spire. The body occupied the outer or living chamber. In the specimen this chamber is now infilled with material identical with the matrix in which the whole fossil is embedded. White crystals of calcite have grown on the walls of the inner chambers, which were gas filled and part of the creature's flotation mechanism. This specimen shows the way in which the spaces left by the disappearance of soft parts may be infilled either by mineral matter or by matrix, so that the originally light and chambered shell has become a heavy and solid object, able to withstand without being crushed the weight of the thousands of feet of rock layers which were deposited on top of it.

The hard parts of organisms are formed of calcium carbonate, calcium phosphate, opaline silica or complex organic compounds known as chitin. These may occur either alone or in combination. Shells, bones, etc., are slightly porous, the pores being infilled by organic tissue. During the process of fossilization, the hard parts may suffer changes, especially those composed of calcium carbonate, for this mineral may have been secreted in either of two forms, calcite or aragonite. Aragonite is much less stable than calcite. During the passage of time shells originally composed of aragonite may be converted to calcite, which has a different crystal structure. As a result of *recrystallization*, with rearrangement of the molecules, a mosaic of interlocking crystals of calcite is produced and the remains of organisms with aragonitic skeletons may be obliterated. This is particularly the case in limestones of considerable geological age. They are likely to contain only those forms whose hard parts were originally composed of calcite.

Calcite and aragonite are both fairly easily soluble in water that is either slightly acidic or slightly alkaline. Ground water percolating through the rocks is therefore liable to dissolve away shells, etc., composed of calcium carbonate or even of silica. If the rock is quite unconsolidated, for instance, a sand, such *solution* will cause the disappearance of any potential fossils, for as the calcite is removed in solution the sand grains left unsupported will slide into the voids. The fact that a sand is now unfossiliferous, does not mean that it never contained fossils – indeed one can sometimes see faint white streaks that are the 'ghosts' of shells almost on the point of disappearance. In a consolidated rock, the cavities formed by the solution of shells are preserved as *moulds*, either external or internal, according to whether the impression on the rock is of the external or internal side of the original. By filling the cavity with suitable material, such as one of the compositions used by dentists for taking impressions of one's teeth and palate, the original form of the shell can be restored in the form of a *cast*. Internal moulds of fossils are particularly useful to the palaeontologist. They show in sharp relief structures which are hidden when the unaltered shell is embedded in rock. Removal of the shell to reveal the details of its interior may require hours of patient and delicate work with a needle or a dental drill or the careful solution of the shell with an acid of such strength that it will dissolve away the shell and not the matrix in which it is preserved. The 'Roach', a quarryman's term for a bed of limestone overlying the valuable building stones in the Isle of Portland in Dorset, is a good example of this type of preservation. The Roach (No. 3) is full of the moulds of two kinds of fossils, the 'screw' of the quarrymen, *Aptyxiella*, a turreted gastropod (p. 153) and the 'horse's head', *Trigonia*, a lamellibranch with a very distinctive dentition (p. 151). Sometimes one finds that solution has affected only the outer part of a block of calcareous sandstone or limestone. When split the outside of the block is seen to be a brownish weathered 'rottenstone' containing only the moulds of fossils, whilst the bluehearted centre of the block is unweathered with the fossils preserved in calcite (No. 4). Such material is clearly excellent for studying both the internal and the external features of fossils.

No. 2 shows a sea-urchin, *Micraster* (p. 162) preserved in flint. The sea urchin has a rigid calcareous shell or test a few millimetres in thickness. As the soft parts within the test decayed, they were replaced by silica gel filtering in through the openings in the test. The internal mould shown on the left was formed. Percolating ground water then dissolved away the calcareous test. In the right-hand photograph, the internal mould has been replaced inside the flint nodule to show the space originally occupied by the test.

Percolating water is a very dilute solution of mineral salts. Under certain conditions one of these salts may be deposited to fill the spaces produced as shell material is dissolved away. In No. 5 the calcareous shell of an ammonite has been lined with iron pyrites, in No. 6 a siliceous sponge *Cephalites* has been replaced by iron oxide and in No. 7 a chitinous graptolite by iron pyrites.

Another possibility is for the mineral salts to be deposited in the pores originally occupied by soft organic material, producing a *petrifaction*. The 'Petrified Forests' of Arizona, Patagonia, etc., provide many exquisite examples of petrification (No. 8), the cells of the fossil wood, cones, etc., having been infilled with silica deposited from the silica-charged waters that are often to be found in volcanic regions.

Plants and those animals with hard parts composed mainly of organic compounds, undergo rather different changes. The more voltatile constituents (hydrogen, oxygen, nitrogen, etc.) are lost and a *carbonaceous film* (Nos. 9 and 10) is left on the bedding planes of the rocks.

Extremely fine-grained deposits may retain the *impressions* made by such soft bodied creatures as jellyfish, whilst on the bedding planes of clays one often finds impressions of the flattened shells of ammonites, etc.

Traces of life of the past are the footprints, trails, coprolites (fossil excreta) and gastroliths (stomach stones of reptiles). *Footprints* (Fig. 1 and page 47) are formed when birds and animals such as dinosaurs, walked across soft, probably wet, mud or sand, which hardened sufficiently to retain the impressions before the next layer was deposited. They are usually easy to recognize and one can deduce from them the gait, length of stride, etc., of their maker. But *trails* and *burrows* may be much more difficult. They are made by organisms, usually invertebrates, walking on or burrowing through the sediment on the floors of bodies of water. At low tide on a modern sea shore one can see many different kinds of trails and burrows made by limpets, worms, etc.

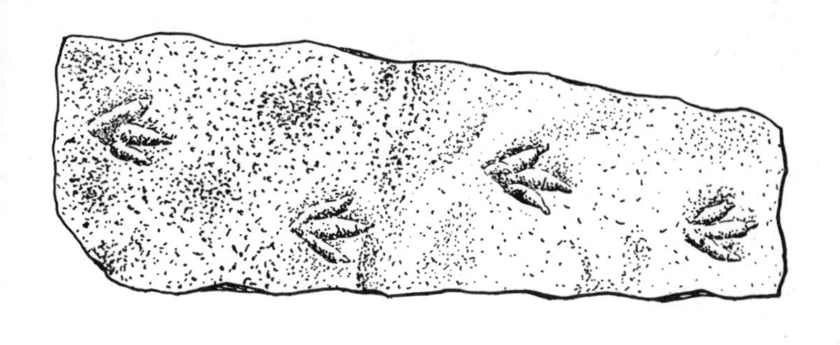

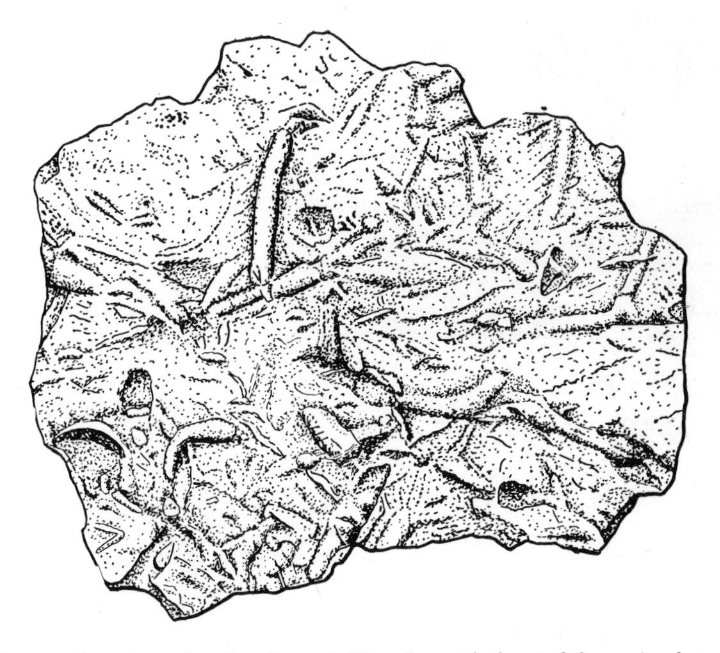

Fig. 1. *Footprints and trails. Above – Bedding plane with three-toed dinosaurian footprints. Below – Bedding plane covered with numerous trails of unknown origin.*

One can relate a definite kind of trail to a particular organism, which may well be soft bodied. But the position is very different when one is dealing with rocks, particularly the older rocks. One often finds peculiar marks on bedding planes. The first question is were they formed by an organism or by a pebble being carried along by bottom currents or some other inorganic cause? They might have been formed by some organism long extinct, so that it is very difficult to deduce the kind of trail it might have made. Within recent years it has become necessay to recognize a group of *trace fossils*, markings which are probably of organic origin (see p. 174).

The Destruction of Fossils

We have already mentioned two ways, by recrystallization and by solution, by which fossils can be destroyed. Another way is by the effects of earth movement or by heat. When rocks are strongly folded, they are metamorphosed. New minerals and new textures are formed. The rocks are reconstituted and all trace of their original nature, including their contained fossils, may be lost. If the metamorphism is not too intense the fossils may still be recognizable, but they will be distorted as the particles which make up the rock are reorientated under the new system of stresses (Nos. 11 and 12). A fossiliferous rock may also be altered by thermal metamorphism. When several cubic miles of molten magma at a temperature of several hundred degrees Centigrade are forced into other rocks, these rocks will be baked. New minerals and textures will be produced and the fossils obliterated. But at any one period of time, only limited regions of the earth's crust were affected by metamorphism, either by mountain building movements or by the extrusion or intrusion of molten magma or a combination of these. In other areas the rocks of the same age were not so deformed. It is from the undeformed or only slightly deformed areas that we obtain our knowledge of the life of the past.

II. THE SCIENTIFIC VALUE OF FOSSILS

The cliffs at Hunstanton in Norfolk show three sharply defined types of rock (No. 13). Whitish chalk resting on a hard red rock containing a few scattered pebbles overlying softer brown pebbly sandstone. The sand was clearly the first layer to be deposited, then the red rock and finally the beds of chalk. The sand is unfossiliferous, but careful search will not only yield fossils from both the red and the white layers, but will also show that the fossils to be found in the two beds are not the same. We can therefore distinguish the beds not only by their obvious differences in rock character (*lithology*), but also by their fossil content.

This cliff section demonstrates several basic geological principles. The Law of Superposition – that in stratified rocks a younger bed overlies an older layer; that different strata contain different assemblages of fossils and, as a corollary, that as the deposition of each rock layer must have taken a certain period of time, the life of the geological past must have changed with the passage of time. Also at Hunstanton the rock layers are not quite horizontal, but are inclined (*dip*) at a gentle angle to the horizontal with the result that at the north-eastern end of the cliff the red rock is at beach level, at the other end (the updip end) well above it. The cliff top is almost level. As one moves downdip younger and younger beds outcrop or cut the ground surface, as shown by the cliff top (Fig. 2).

If one follows the outcrop of the basal beds of the Chalk southwards from Hunstanton towards Cambridge, one finds that wherever they are exposed in quarries, cuttings, etc., in the north chalk rests on the red rock, but further southwards the red rock gradually passes into a blue clay, known as the Gault. The Gault is overlain by chalk and is underlain by sands. It is in the middle of a 'geological sandwich' similar to that exposed at Hunstanton. Fossils, identical to those found in the red rock of Hunstanton can be obtained from the Gault. The lateral change from sand to clay must reflect a change in the nature of the bottom conditions of the sea

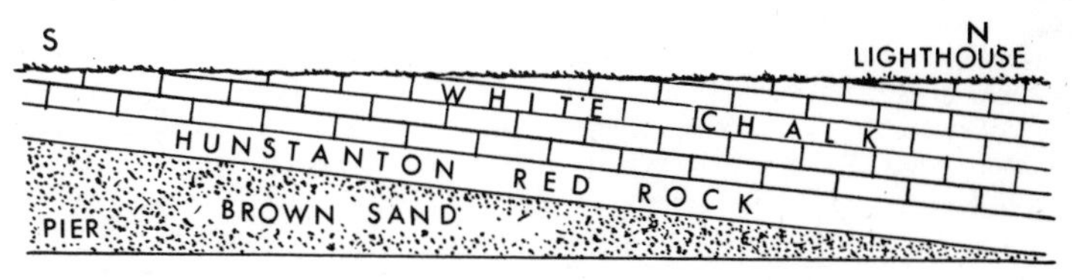

Fig. 2. *Section in the cliffs at Hunstanton, Norfolk.*

in which both the Gault and the red rock were laid down.

The methods used in Norfolk can be applied anywhere to arranging stratified rocks in their order of deposition. So far we have only been considering a few score feet of gently dipping rocks, outcropping more or less along a straight line. When these methods are applied to larger regions, such as the whole of England, one can recognize a great number of rock groups or formations, some such as the Chalk thick and traceable over wide areas, others thinner, localized and passing laterally into beds of a different lithology, e.g. the Hunstanton Red Rock and the Gault Clay. When all the formations are arranged in their order of deposition, we have built up a **Stratigraphical Table**, but in this form it would consist of hundreds of units. Therefore the Stratigraphical Table is divided first into the major units or *Systems*, shown in Table I. In any area each System may be represented by a number of formations. The three formations which we have been considering so far represent only a part of the Cretaceous System. The few feet of Chalk at Hunstanton dip eastwards and are overlain in eastern Norfolk by about fifteen hundred feet of younger Chalk, whilst in south-east England the Gault is underlain by two thousand feet of older sands and clays, forming the lower (older) part of the Cretaceous System.

The nomenclature of the geological systems is rather irrational. It dates from over a hundred years ago, when the main subdivisions into which the stratified rocks could be subdivided were being hammered out. Some of the names – Cambrian (Cambria the Roman name for Wales), Permian (after a town on the River Volga in Russia), Jurassic (the Jura Mountains), Devonian, are from the type areas where a particular system was first recognized; others, e.g. Cretaceous (*creta*, the Latin name for chalk), Carboniferous (the coal-bearing rocks of the British Isles), Triassic (from the type area of north Germany, where these beds show a three-fold division into sand–limestone–marl), refer to distinctive lithological features. The Ordovician and Silurian Systems were named after Celtic tribes that inhabited the type areas in the Welsh Borderlands. Most logical are the names for the youngest systems; Palaeocene (Greek *ancient recent*), Eocene (*dawn of the recent*), Oligocene (*little recent*), Miocene (*less recent*), Pliocene (*more recent*), Pleistocene (*most recent*), Holocene (*wholly recent*), for these were originally based on the comparison of the fossil content of the rocks of each system with the recent fauna that now inhabits the Earth. The older the system, the smaller its percentage of recent forms, or conversely the greater the percentage of forms that had become extinct during the passage of time.

In its type area a geological system is represented by rocks many thousands of feet in thickness. These rocks were deposited during a certain *period* of time, so in a chronological sense we speak of the Silurian Period, etc. Periods are grouped together (see Table I) into larger time units known as *eras*. Eras are named in two different ways, first Primary, Secondary, Tertiary and Quaternary, the order of deposition; secondly according to the main features of their fossil content, Palaeozoic (*ancient life*), Mesozoic (*middle life*), Cainozoic (*recent life*). The terms Primary and Secondary have long fallen into disfavour, whilst Tertiary and Quaternary are used by most British geologists, but many foreign and American geologists prefer Cainozoic or Caenozoic.

Terms that have been introduced recently are Phanerozoic (*clear evidence of life*) and pre-Phanerozoic. As shown in the Table, it is in the rocks of the Cambrian System that fossils first appear in any degree of abundance. The underlying Pre-Cambrian rocks were formerly thought to be unfossiliferous, but this is not now the case. Recognizable fossils have now been found in the Pre-Cambrian rocks at a number of localities throughout the world. The great majority of the organisms of pre-Phanerozoic times must have been without hard parts and therefore the chances of their preservation in

The Stratigraphical Table

ERA		PERIOD	DURATION	BEGINNING OF PERIOD FROM PRESENT	DISTINCTIVE FORMS OF LIFE
Quaternary		Holocene		0.01	Man
		Pleistocene	c.2	c.2	
Tertiary		Pliocene	5	7	Modernized mammals
		Miocene	19	26	
		Oligocene	12	38	Archaic mammals
		Eocene	27	65	
Mesozoic		Cretaceous	70	135	Age of reptiles, ammonites and belemnites. In the Cretaceous angiosperms supplanted gymnosperms
		Jurassic	40	195	
		Triassic	30	225	
Palaeozoic	Upper	Permian	45	280	Pteridosperms dominant.
		Carboniferous	65	345	Rise of amphibians
		Devonian	50	395	Rise of fish and of primitive land plants
	Lower	Silurian	35	430	Age of graptolites, trilobites and many other invertebrates
		Ordovician	70	500	
		Cambrian	70	570	Fossil invertebrates become common
Pre-Cambrian or Pre-Phanerozoic				4,600	Extremely rare traces of life, algae, etc., to at least 1,700 million years ago

Note all ages are given in units of 1 million years.

identifiable form are excessively small.

Within recent years it has become possible to estimate with considerable precision the actual duration of the various periods. For brief details of the method see p. 144 of the author's *Minerals and Rocks in Colour*.

UNCONFORMITY AND MOUNTAIN BUILDING

In any one area, even an area as large as the British Isles, the full quota of the geological systems is not present. No. 14 shows a section at Portishead on the southern shore of the Bristol Channel. A bouldery deposit with a horizontal base overlies massively bedded brown sandstones dipping at about 25° to the left of the picture. The upper bed is of Triassic, the lower sandstones are of Devonian age. The pebbles and boulders in the upper bed are similar in composition to the underlying sandstone. After their deposition the Devonian sandstones must have been tilted, uplifted and finally eroded to provide the material for the Triassic boulder bed. The passage of more than 100 million years (the duration of the Carboniferous and Permian periods) is not represented here by deposits. At Hunstanton the three formations present are *conformable*, that is they are parallel to each other. At Portishead the Triassic beds rest *unconformably* on the Devonian strata and the two rock-groups are separated by a clearly marked surface of unconformity.

The unconformity at Portishead was caused by an *orogenic episode*, a period of mountain building when the rocks of a part of the Earth's crust were strongly folded, intruded by granites and finally uplifted. Three such episodes have effected the Phanerozoic rocks of western Europe – the Caledonian orogeny towards the end of the Lower Palaeozoic times, the Variscan (often called the Armorican) orogeny in late Palaeozoic times and the Alpine orogeny in mid-Tertiary times. As its name suggests the maximum stresses of the Caledonian orogeny were felt in the Highlands of Scotland. The centre of the Variscan mountain belt lay through France, Belgium, central and southern Germany, whilst as its name implies, the Tertiary orogeny produced the Pyrenees, Alps, Carpathians and other of the young mountain chains of Europe.

Other parts of the world have been effected by different orogenic episodes, but the result has always been the same, to produce gaps in the stratigraphical record with the beds laid down after a particular period of folding resting unconformably on tilted, folded and often metamorphosed rocks.

FOSSILS AS TIME INDICATORS

It will be seen from the preceding pages that the rock layers which make up the Earth's crust are arranged in an orderly manner, though in regions which have been strongly effected by orogenic movements, the beds have been greatly disturbed and their relationships may be very difficult to elucidate. For a wide variety of purposes, ranging from palaeogeography, the reconstruction of geographical conditions at a definite interval of time in the geological past, to the location of suitable structures, such as unconformities, faults, folds, etc., in which mineral oil or gas may be trapped, fossils are one of the most useful tools available to the geologist.

The corollary to William Smith's dictum (p. 9) that 'strata can be identified by their contained fossils' is that a particular kind of fossil organism only lived during a certain period of the geological past and hence is only to be found in rocks of that age. Some organisms have a very long time range. The brachiopod *Lingulella* (p. 133 and No. 48) found in the Cambrian rocks of North Wales is almost identical with *Lingula* living today off the coasts of Japan. But this is exceptional, most fossils are much more restricted in their range. The sea urchin *Micraster* (p. 162 and No. 130) is only to be found throughout a thickness of a few hundred feet of Chalk, whilst the crinoid *Marsupites testudinarius* (p. 160 and Fig. 29) only occurs in a layer of Chalk, little more than a score of feet in thickness.

One can therefore subdivide the rocks of each System into *zones*, each of which is characterized by a particular assemblage of fossils, of these one is selected as the *zonal index* and the zone is named after it. Zones are grouped together into large units, known as *stages*. For instance the zone of *Marsupites testudinarius* is a part of the Senonian Stage of the Cretaceous System. To identify a zone, one does not have necessarily to find the zonal index; it is sufficient to find sufficient members of the assemblage peculiar to that zone. Each zone must represent a particular period of time. How long a period we do not know precisely as yet, though it is probable that in the future with the development of radiometric dating of rocks we may be able to define precisely the duration of at least some zones. At the moment, all that can be said is that the rocks of a particular zone were deposited during a period of time measurable in many thousands, and in the Older Palaeozoic rocks, in millions, of years. It is reasonable to regard all the rocks belonging to a particular zone as approximately contemporaneous.

Not all fossils are suitable for zoning the rocks. Clearly the first factor is time range. The narrower the time range of a fossil, the greater its value. Secondly the fossil must be reasonably common and preferably easily recognizable. Owing to their distinctive ornamentation, even broken fragments of *Marsupites* and *Micraster* are identifiable, but at other levels it may be necessary to find almost complete specimens before identification can be made with certainty. Thirdly zone fossils must have a wide distribution geographically. This is much more likely to be the case if they are not *benthonic* (bottom living forms). Most benthonic forms originate as larvae, drifted about by currents. The larvae finally settle on the sea floor and only if conditions are suitable can they develop into adults. Their distribution is therefore controlled by the ecological conditions, sometimes distinctly specialized conditions, of the sea floor. On the other hand, forms which are either *nektonic* (swimming) or *planktonic* (drifting) throughout their life are much more widely distributed, for conditions change much less sharply in the waters of the oceans than on the sea floor. Furthermore *pelagic* forms (planktonic and nektonic) may be distributed after death by currents transporting the slowly sinking dead organisms. This cannot occur with benthonic forms. It therefore follows that ideally the fossils used for zoning the rocks should be pelagic forms with a skeleton, which is likely to be resistant to the hazards of fossilization, e.g. *Marsupites*, a nektonic sea-lily (crinoid). But at many horizons, suitable pelagic forms do not occur sufficiently commonly and then one has to use benthonic forms, such as the burrowing sea-urchin *Micraster*.

We have already described (p. 15) the lateral passage of the Hunstanton Red Rock into the clays of the Gault. The two formations are contemporaneous for they yield the same zonal fossils. When one traces the Gault Clay south-westwards along the foot of the Chiltern Hills and the chalklands of Wessex, it passes laterally in Wiltshire and north Dorset into greenish sands and sandstones, the Upper Greensand. This bed yields the same pelagic fossils as the Hunstanton Red Rock and the Gault and, in addition sponges, thick-shelled lamellibranchs and corals, a benthonic assemblage that lived in much shallower water than the benthonic forms that are found in the East Anglian outcrops.

FACIES AND FACIES FOSSILS

Facies is a term much used in modern geology. The lithology and fossil content of beds is determined by the environment in which they were deposited. The Hunstanton Red Rock, the Gault and the Upper Greensand are different facies, the calcareous, argillaceous (clayey) and arenaceous (sandy) facies, of part of one stage, the Albian, of the Cretaceous System of Britain.

So far we have been considering one rather narrow time-interval with the same pelagic zone fossils (certain ammonites) present in the rocks of all three facies and hence proving their contemporaneity.

But at other levels it may be necessary to use different zone fossils for different facies. The rocks of the Ordovician System are many thousands of feet in thickness. In western and central Wales the Ordovician rocks are mainly compacted clays yielding graptolites, rather fragile pelagic forms (p. 172). The Ordovician rocks can be divided into a number of stages and each stage into a number of zones, each characterized by distinctive graptolites. In Shropshire (Fig. 3) the Ordovician rocks are of a different facies, mainly sandstones with some thin limestones. The beds were laid down in much shallower water than those of the graptolitic facies. They yield a benthonic fauna of trilobites and brachiopods and can be subdivided into stages and zones each characterized by a distinctive trilobite–brachiopod assemblage. Graptolites are only found in the rocks of the sandy facies in occasional shaley layers. The shales were laid down in much quieter conditions than the sands and hence preserve the fragile graptolites, which were ground to pieces at other horizons. Fortunately

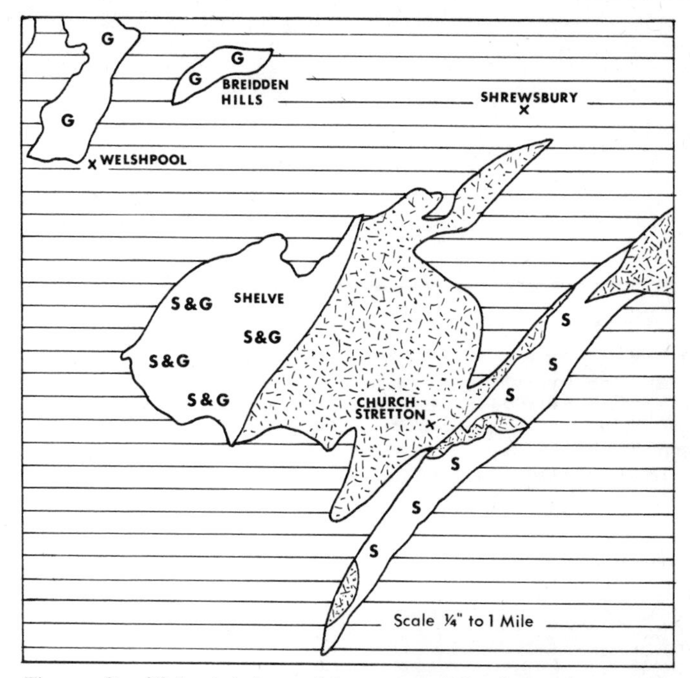

Fig. 3. *Simplified geological map of the area around Church Stretton, Shropshire.*

Horizontal Ruling	*Beds younger than the Ordovician*
White	*Outcrop of the Ordovician rocks*
S – Sandy facies	*G – Graptolitic facies*
Stippled	*Beds older than the Ordovician*

there are sufficient graptolite-bearing shale horizons in the Orodovician rocks of the sandy facies, for the two Zonal Schemes, the one based on graptolites, the other on trilobites and brachiopods, to be correlated (Fig. 4) and also to enable the original distribution of these three different groups of fossils (Fig. 5) to be deduced.

Many fossils are restricted to beds of a particular lithology, indicating that their conditions of life were specialized. For example, in the Upper Jurassic rocks of the French Alps there are developments of thick limestones yielding many corals, sea-urchins and thick-shelled lamellibranchs. The type of faunal assemblage that is found today around coral reefs. The detailed characters of the limestones show that these beds were laid down as part of a reef belt. These reef-limestones pass laterally into thin bedded limestones and marls yielding ammonites.

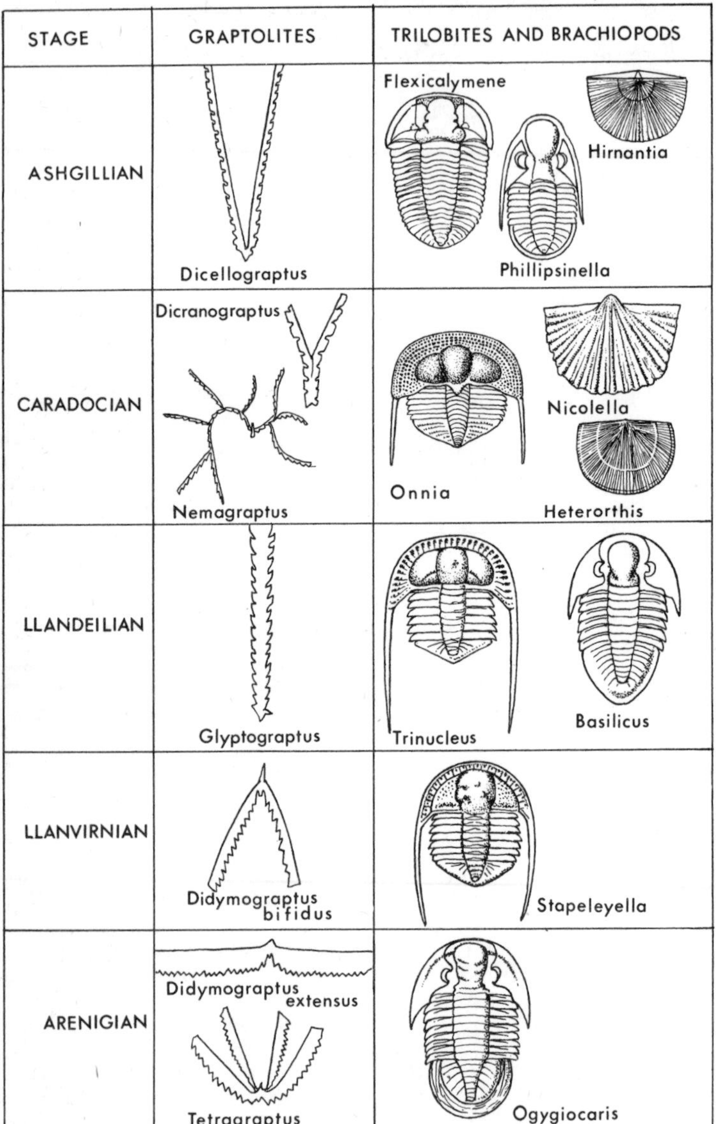

STAGE	GRAPTOLITES	TRILOBITES AND BRACHIOPODS
ASHGILLIAN	Dicellograptus	Flexicalymene Hirnantia Phillipsinella
CARADOCIAN	Dicranograptus Nemagraptus	Nicolella Onnia Heterorthis
LLANDEILIAN	Glyptograptus	Basilicus Trinucleus
LLANVIRNIAN	Didymograptus bifidus	Stapeleyella
ARENIGIAN	Didymograptus extensus Tetragraptus	Ogygiocaris

Fig. 4. *The Ordovician System; its stages and a selection of the fossils used for zonal purposes.*

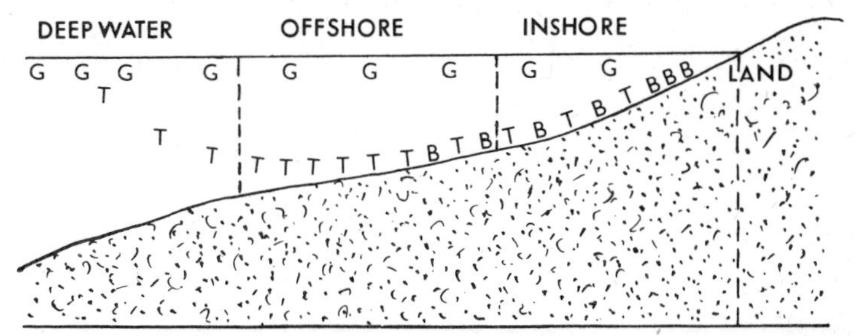

Fig. 5. *Diagrammatic N.W.–S.E. section across Figure 3 showing the ecological zones inhabited by B – Brachiopods; G – Graptolites; T – Trilobites.*

Ammonites are excellent zone fossils, for they were pelagic and of strictly limited time-range. The relationships between the ammonite zones and the reef-limestones are shown in Fig. 6. Conditions suitable for reef-building organisms to flourish must have developed first around Besançon, and then spread slowly south-eastwards so that the reef belt is *diachronous*, that is it cuts across the time-planes based on the ammonite zones. Such diachronous relationships are always liable to occur in sedimentary rocks and therefore it is most important to distinguish between those fossils that are of true zonal value and those that are facies fossils – of long time-range and restricted to specialized environments. If this is not done erroneous conclusions may well be drawn as to the relationships of beds and hence of the palaeogeography of the past.

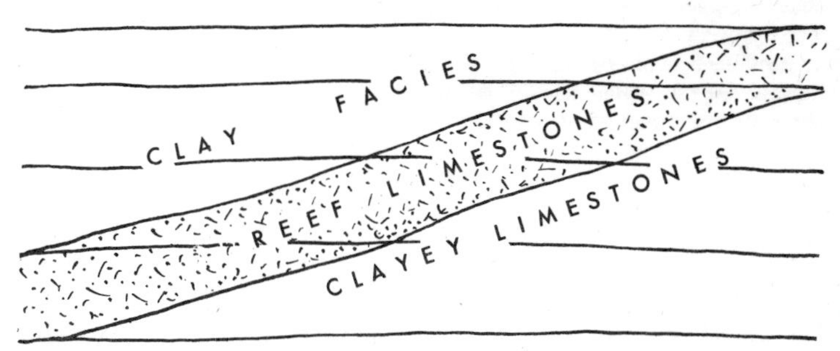

Fig. 6. *Diachronism. The facies belts cut across the horizontal lines, which represent time planes.*

THE VALUE OF ZONE FOSSILS

Zone fossils are often of the greatest value to the geologist who is concerned with subdividing stratified rocks, so that he can trace time-units

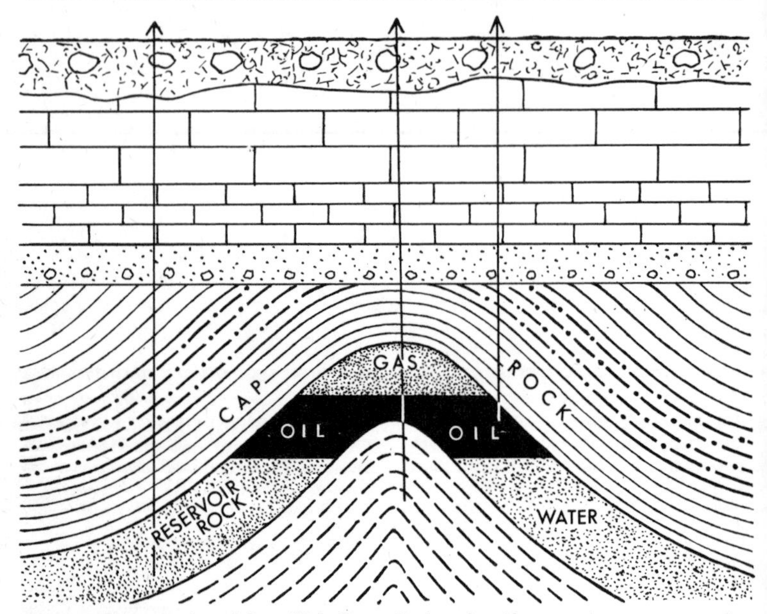

Fig. 7. *Interpretation of Bore Hole Data. Borings for oil penetrating an upper series resting unconformably on a strongly folded oil-bearing lower series of rocks.*

through them, and correlate rocks of the same age but of differing lithology. The oil-field geologist is concerned with locating structures in which oil may accumulate. The boring may penetrate an unconformity. If the beds beneath the unconformity are fossiliferous, he can determine their age. If the same unconformity has been penetrated in other borings, he can deduce the structure of rocks, perhaps a mile beneath the earth's surface and completely hidden by their unconformable cover (Fig. 7). Geophysical methods of investigation can also provide information as to the disposition of rock layers at considerable depth, but they only record the presence of rock layers with distinctive physical properties. The only way to determine the age of a particular layer is to examine samples obtained by the drill. To obtain as complete a picture as possible of the nature, age and structure of rocks deep beneath the earth's surface, it is necessary for the sub-surface geologist to combine all the information that can be obtained by the geophysicist and the micropalaeontologist. How vital is the role of the micropalaeontologist is shown by the numbers employed by the oil industry. In the early days of oil exploration, the emphasis was on finding suitable structures, more recently the outlook has widened. Much greater attention is now being paid to deducing the conditions under which rocks were laid down. An area may contain suitable structures, but they will be barren if conditions were never suitable for the formation of oil. In both academic and industrial geology much greater attention is now being paid to the construction of detailed palaeogeographical maps for strictly limited time-intervals and these can only be defined if satisfactory zone fossils are available.

LIFE AND DEATH ASSEMBLAGES

It is very important when studying a fossil assemblage to determine whether it has been preserved in the place where the organisms lived or whether it consists of organisms that have been drifted or otherwise collected together after death. A life assemblage (No. 16) would contain a mixture of juvenile and adult individuals, burrowing forms would be in the position of life, at right angles to bedding planes, the root systems of plants would be preserved and so on. A death assemblage, on the other hand, would show definite signs of sorting, its members might well be aligned by current action, they would be broken and show signs of wear, it might contain a mixture of marine and terrestrial forms or of marine forms that lived in different habitats. An extreme example of a death assemblage is a bone bed (No. 15) consisting of heavily worn and phosphatized material. Extreme cases such as those given above are easy to recognize, but there are many more subtle ones where it is much more difficult to be sure how much disturbance there has been after death.

FOSSILS AS INDICATORS OF CLIMATIC CONDITIONS

Modern plants and animals are sensitive, sometimes very sensitive, to the climatic conditions under which they can survive. If a fossil life assemblage includes organisms that are not extinct, then one has evidence as to climatic conditions of the past. Insects and plants are very useful in this respect, especially in the study of the Quaternary deposits. The clayey deposits laid down in ponds on land or in shallow estuaries and lagoons off-shore will contain the pollen and spores of plants that grew round them and the harder parts of the insects that lived in or flew over the waters. During the Quaternary Era there was rapid climatic variation due to a sequence of glacial periods separated by interglacial periods. The climatic conditions that prevailed whilst a particular muddy layer was being deposited can be deduced within narrow limits by the determination of its content of pollen grains and insect fragments and then the consideration of the conditions under which such forms live today.

Clearly the older the deposit with which one is dealing, the less exact will be this type of information, for the higher will be the percentage of extinct forms. With marine deposits, containing a mixture of benthonic and pelagic forms, climatic control will be less significant and hence more difficult to recognize. Moreover there is clear evidence that for much of the geological past the climatic belts were not so varied or sharply defined as they are at present.

FOSSILS AND EVOLUTION

The pioneer geologists, such as William Smith and Cuvier and their successors who hammered out the order of the systems of the Stratigraphical Table, showed that the successive group of rocks contained different assemblages of fossils, but there was much puzzlement as to the relationship between the faunas. Much geological thought in the early part of the last century was strongly biased towards catastrophism. Valleys were regarded as having been opened by 'convulsions of nature', great floods were readily invoked to account for otherwise puzzling pebbly and bouldery deposits. In the same way many thought that each fauna was widely distributed for a time and then was extinguished by some great catastrophe. A new fauna was created or immigrated into the area only to be wiped out in its turn. Others appreciated more clearly the immensity of geological time, that rivers in the course of time erode the valleys in which they flow, that bouldery deposits can be formed in a variety of ways and that whilst in some areas rocks containing say a Lower Silurian fauna rest with an unconformable junction on beds yielding Middle Ordovician fossils, yet in other areas there is conformity between the rocks of the two systems. Indeed in such areas careful bed-by-bed collecting has shown that there is a gradual change through 'passage beds' from strata yielding an undoubted Upper Ordovician fauna to rocks containing a typical Lower Silurian assemblage.

Then in 1859 appeared Charles Darwin's great *Origin of Species* in which he argued that species had arisen from pre-existing species by a process of descent with modifications. Earlier workers such as Lamarck (1744–1829) had suggested the transformation of species, but they had not been able to bring forward the wealth of supporting detail that Darwin did, nor could they suggest such a convincing mechanism for evolutionary change. Darwin's mechanism was the inheritance of those variants most favoured by natural selection acting on the considerable variation in detail shown by organisms belonging to the same species. Those organisms which are best adapted to their particular environment must survive and breed at the expense of those which are less well adapted. The reality of evolutionary change soon became generally accepted after the publication of Darwin's great work. Whilst there followed much controversy as to the precise mechanism by which evolution has operated, the majority of palaeontologists, neontologists and geneticists now favour a modified form of natural selection.

In Chapter IV we shall trace in outline the evolutionary history of most of the main groups of fossil organisms. Owing to the chances of preservation and of discovery there are many gaps in the record, but the main trend or

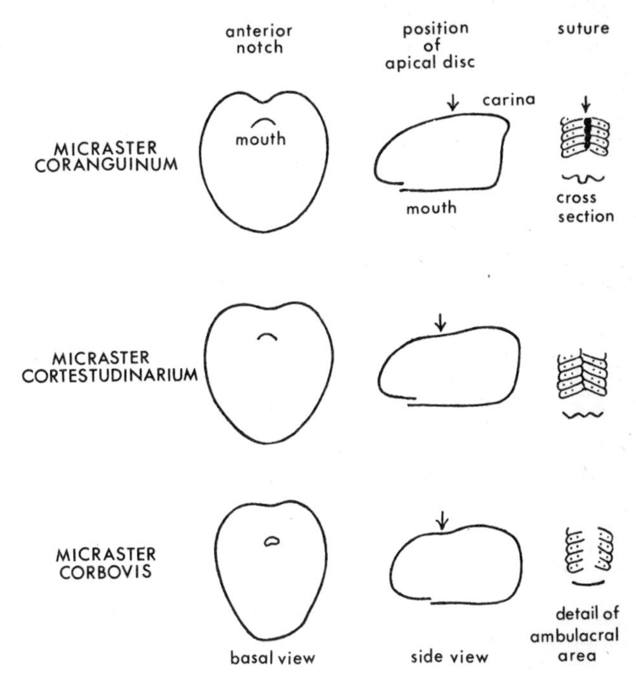

Fig. 8. *Members of the* Micraster *lineage, showing the chief variations in morphologica features.*

trends are usually clear. Exceptionally one can find examples of **evolutionary series** showing the transformation of one species to another. Such a series is shown by those common fossils of the Upper Chalk, the Micrasters. Throughout a thickness of about 400 feet of pure soft white limestone, these sea-urchins show regular changes in a number of minor morphological features, so that *Micraster corbovis* clearly differs from the much later *Micraster coranguinum* (Fig. 8). These changes do not all take place at exactly the same rate, but there is sufficient general correspondence for it to be possible to divide the *Micraster* lineage into a number of arbitrarily defined species, each of which is restricted to a certain thickness or zone of the Upper Chalk. Therefore such an evolutionary series is of great practical value as it provides an accurate scale for the detailed zoning of beds.

III. THE OCCURRENCE OF FOSSILS

The nature of the rocks present in any area is determined first by the conditions under which the rocks were formed and secondly by any changes to which they may have been subsequently subjected. Another factor governing the fossil content, if any, of the rocks is their geological age. The fauna of a marine mud of Lower Palaeozoic age will clearly be very different from that found in a marine mud of Mesozoic age. During the passage of several hundreds of millions of years, the majority of the groups of organisms that occurred in the Lower Palaeozoic rocks will have passed into extinction and different groups will have arisen to take their place.

In the next chapter, we shall be describing in turn each of the chief groups of fossil organisms, showing how they functioned, the particular conditions under which they lived and the changes that occurred within each group during the passage of time.

The areas in which one may hope to find a certain fossil have been determined by the events of the geological past. In the first place, its original distribution over the earth's surface was controlled by the geographical conditions of the time at which it lived. These determined how widespread was its special habitat. But later events have also played a very important part. The beds containing these particular fossils may have been buried so deeply beneath later deposits, that they do not now outcrop on the surface and can only be reached by the drill; these rocks may have been so deformed by subsequent earth movements that their fossil content has been obliterated; these strata may have been eroded away at some later time, so that the only chance of finding the fossils which they originally contained, is if these fossils have survived as derived fossils in deposits of later age; or the fossiliferous rocks may have escaped all these possibilities and outcrop on the existing landsurface, though it is extremely unlikely that their present outcrop is nearly as extensive as the area over which these particular beds were originally deposited.

Tracing the history of a group of fossils, therefore, involves more than the evolutionary changes that occur with the passage of time. One must also consider why at one period the members of the group are only to be found in certain areas, whilst at another period their distribution may be very different.

Much repetition in the next chapter will be saved, if we first consider briefly the outcrop pattern of the various major rock groups on the landsurface of western Europe and then go on to sketch in outline, with the help of palaeogeographical maps, the long sequence of events that changed so often during the past 600 million years, the form and appearance of that part of the earth's crust which now forms western Europe.

THE GEOLOGICAL MAP OF WESTERN EUROPE

Scandinavia and Finland are mainly an area of metamorphic and igneous rocks; in the Baltic Shield (Map 8) of Pre-Cambrian age, but in western Norway partly of Pre-Cambrian and partly of Lower Palaeozoic age. The Lower Palaeozoic strata of Western Norway were so deformed during the Caledonian earth movements that they yield fossils at very few localities and those that can be found are but poorly preserved. On the margins of the Baltic Shield in Esthonia, in the island of Gotland, in southern Sweden and around Oslo in southern Norway, the Pre-Cambrian rocks are overlain unconformably by richly fossiliferous shales and limestones of Lower Palaeozoic age, whilst in the extreme south of Sweden in Scania there is some development of Jurassic and Cretaceous beds. Finally much of Scandinavia is thickly mantled by the boulder clay and gravels laid down during the Pleistocene glaciations, whilst raised beaches, often fossiliferous, fringe the coasts of the Baltic Sea. Clearly there are big gaps in the record, for virtually no rocks of Upper Palaeozoic age are present, whilst the succession of the Mesozoic and the Tertiary strata is very incomplete.

The record in **Denmark** has even more gaps. The only 'solid' rocks outcropping beneath the widespread glacial 'drifts' of Pleistocene age are the chalks of Zealand in the east and the Lower Tertiary clays of Jutland in the west. These beds are not only richly fossiliferous, but also include the unique Danian Chalk of either latest Cretaceous or earliest Tertiary age, for the precise age of the Danian Stage is still in dispute.

The widespread sandy 'drift' deposits of **North Germany** overlie a thick succession of marine clays of Tertiary age, poorly exposed at the surface, but well known from the numerous borings for oil. Further southwards in the more hilly country towards Hanover and Osnabruck there is a complete succession of gently dipping and undeformed rocks of Mesozoic age extending southwards into Bavaria, where the Jurassic limestones form the great scarp of the Franconian Jura. Eastwards the Mesozoic beds abut against the metamorphic and igneous rocks of the Bohemian massif. Around Prague the Pre-Cambrian rocks of the massif are overlain unconformably by undeformed Lower Palaeozoic and Cretaceous rocks. Magnificent faunas have been obtained from the Cambrian rocks of Bohemia. The Mesozoic strata of Hesse overlie unconformably the Devonian and Carboniferous rocks of the Rhenish Schiefergebirge. The Palaeozoic beds were folded during the Variscan orogeny, but not too severely, so that they still contain a rich fauna. The Upper Palaeozoic rocks of the Rhenish Schiefergebirge continue eastwards into the Ardennes of southern Belgium but there is a facies change, for the Devonian and Carboniferous rocks of the Rhenish Schiefergebirge are mainly shales, those of the Ardennes mainly limestones. The Ardennes and the Rhenish Schiefergebirge are the European type area for the study of the marine faunas of Devonian age. To the north of both the Rhenish Schiefergebirge and the Ardennes are the Coal Measures of the Ruhr and the Namur Basins, overlain unconformably by Cretaceous and Tertiary rocks dipping gently northwards towards the North Sea. The Devonian rocks of the Ardennes in their turn rest unconformably, around Spa and Stavelot, on steeply dipping slates and quartzites. These have yielded a few poorly preserved Lower Palaeozoic fossils, but as in western Norway, the fossil evidence is so meagre that these beds are often referred to as the Cambro-Silurian.

The dominant feature of northern **France** is the Paris Basin with the almost horizontal Lower Tertiary beds in the centre surrounded by concentric outcrops of Cretaceous and Jurassic rocks. To the west the Mesozoic beds rest unconformably on the strongly folded Palaeozoic rocks of Normandy and the Brittany Peninsula. The southern boundary of the Paris Basin is formed by the Massif Central, where Upper Tertiary volcanics and sediments overlie strongly folded and metamorphosed Palaeozoic rocks. To the west of the Massif Central lies the Aquitaine Basin infilled with Mesozoic and Tertiary rocks, to the east the Rhône Valley.

In the parts of Europe that we have been describing so far, the Mesozoic and early Tertiary rocks are but little disturbed, but to the south lie the *Young Mountain Chains of Europe*, the Pyrenees, the Alps and then further eastwards the Carpathians, all formed by the Alpine orogeny. In the heart of the chains the Mesozoic rocks have been so strongly metamorphosed that recognizable fossils are extremely rare, but on the northern sides of the chains the folding is less intense. To the north of the folds proper lies the Causse and the French and Swiss Jura, with thick developments of biohermal limestones of Mesozoic age and further east in northern Switzerland, extreme southern Germany and northern Austria, is the Molasse Basin, infilled with undeformed and richly fossiliferous Upper Tertiary deposits.

In no other European area of comparable size are there such a variety of geological formations as in the **British Isles**. Scotland north of the Midland Valley is composed of folded and metamorphosed rocks of Pre-Cambrian and probably earliest Palaeozoic age. These are overlain unconformably on the east by continental deposits, the Old Red Sandstone of Devonian age and the New Red Sandstone of Permo-Triassic age, and on the west coast by thin and rather scattered outcrops of Lower Palaeozoic and Mesozoic rocks. The Midland Valley of Scotland is floored by Upper Palaeozoic strata resting unconformably on small inliers of folded Lower Palaeozoic rocks, whilst Lower Palaeozoic beds outcrop extensively across the Southern Uplands. The Upper Palaeozoic rocks are but slightly folded, the Lower Palaeozoic rocks much more intensively. The same relations hold in most of Ireland and in north Wales, but in the southern part of Wales there is a conformable passage from the Silurian strata up into the Old Red Sandstone of Devonian age. We are passing southwards into areas where the effects of the Caledonian orogeny were much less intense. But in extreme southern Ireland and in the Cornubian Peninsula, it is the Upper Palaeozoic rocks that were strongly folded and often metamorphosed by the Armorican orogeny. These beds are overlain unconformably, as at Portishead (No. 14), by Mesozoic rocks. In northern England there are wide outcrops of gently dipping Carboniferous rocks resting, on the west, with marked unconformity on the Lower Palaeozoic beds of the Lake District. To the east Carboniferous rocks are overlain with no great angular discordance by strata of Permian and Triassic age, beds mainly of continental origin, the New Red Sandstone, which outcrops over much of the Midlands. From the coast of north-east Yorkshire to Dorset there extends a great pile of conformable strata dipping gently south-eastward and of Jurassic and Cretaceous age. In south-east England the beds were gently folded during the Alpine Movements, so that the Cretaceous rocks of the Weald now separate the London and Hampshire Basins infilled with Lower Tertiary strata. It is only the Miocene System that is unrepresented, for in East Anglia rocks of Pliocene and of the early Pleistocene age overlie, with slight unconformity, the Chalk and Lower Eocene beds. Finally we must mention the extensive spreads of 'drift' deposits which often hide the 'solid' rocks especially in areas of low relief. To the north of a line running from just north of the Thames to the Severn, these drift deposits are mainly boulder clay and outwash gravels from the Pleistocene ice sheets, to the south of this, ancient river gravels and ill-sorted deposits laid down under freeze–thaw conditions on the tundra that lay beyond the ice sheets.

THE CHANGING GEOGRAPHY OF EUROPE DURING PHANEROZOIC TIMES

The *Lower Palaeozoic rocks* of Scotland, northern England, Wales and Ireland show two distinct facies, the graptolitic and the sandy (see p. 20). These beds were laid down in a *geosyncline*, that is a belt of the earth's crust which subsided during a long period of geological time, so that it became infilled with a pile of sediments several miles in thickness. The graptolitic facies indicates the axial region of the geosyncline, the sandy facies the margins. This geosyncline trended south-west–north-east across Wales, the Lake District and southern Scotland. It was bounded to the north-west by a stable (non-subsiding) foreland formed by the Pre-Cambrian rocks of north-west Scotland, on which rested a thin shelf development of sandstones and limestones of Cambrian and earliest Ordovician age. The south-eastern margin of the geosyncline lies hidden beneath the Upper Palaeozoic and Mesozoic rocks of England, but for long periods it cannot have been far to the east of the existing most easterly outcrops of Lower Palaeozoic beds. These were clearly deposited in shallow water with shelf limestones particularly well developed in the rocks of Silurian age.

The Lower Palaeozoic rocks of western Norway must have been deposited in the north-eastern continuation of the same geosyncline, with the Pre-Cambrian rocks of the Baltic Shield forming a stable area to the south-east. In southern Norway, southern Sweden and Esthonia, shelf deposits including reef-limestones, were laid down on the submerged margins of the Baltic Shield.

The outcrops of Lower Palaeozoic rocks in the other parts of Europe are of small extent and widely scattered so that it is difficult to reconstruct a palaeogeographical picture. A possible interpretation is shown in Map 1, but it is probably an over-simplification. The strong similarities between the marine faunas yielded by the Lower Palaeozoic rocks of England and Wales, southern Scandinavia and Bohemia show that these areas cannot have been separated by permanent barriers for these would have prevented migration. On the other hand there are considerable differences in detail between the faunas yielded by the Lower Palaeozoic rocks of north-west Scotland and Wales. The deep waters of the axial portion of the geosyncline must have been a barrier to the migration of forms such as trilobites (p. 167) and brachiopods (p. 133), which inhabited the shallower margins of the geosyncline.

Geosynclines are belts of weakness in the Earth's crust. When a geosyncline has been filled with sediment, then follows the phase of lateral compression. The sediments are folded, metamorphosed, intruded by granites and finally uplifted to form a mountain range. The process is usually not quite as simple and clear cut as this. During the sedimentary phase there are very liable to be pulses of crustal unrest causing the extrusion of volcanic rocks and some degree of folding. As a result the succession of beds, particularly in the marginal parts of the geosyncline, will not be complete and conformable, but will be interrupted by minor unconformities.

This is what happened in the British Isles and western Norway during the latter part of the Lower Palaeozoic Era. The final crescendo of the *Caledonian Orogeny or Mountain Building Episode* took place some 400 million years ago towards the close of the Silurian Period, but there were many preliminary tremors. The geography of north-western Europe was profoundly changed. Great south-west–north-east trending mountain chains were formed stretching from the west coast of Norway to north Scotland. In the area of maximum crustal stress the Lower Palaeozoic rocks were considerably metamorphosed. In southern Scotland, the Lake District and north Wales, the beds are less deformed, though in certain areas, such as the Welsh Slate Belt, around Llanberis, the muds have been changed into slates with the development of new planes of splitting, cleavage planes, at a high angle to the original bedding. The mountains must have merged south-eastwards into foothills lying across south Wales and midland England. It is impossible to determine how great was the contraction of the crust caused by the Caledonian orogeny. All one can be sure of is that in Lower Palaeozoic times, north Scotland must have been considerably more distant from south Wales than it is today.

At the beginning of the Upper Palaeozoic Era a new geosyncline, the *Variscan geosyncline*, came into being with its axial region running approximately east–west from Brittany across the south of the Ardennes to the Rhineland and the Harz Mountains. To the north lay the British–Scandinavian land mass formed by the welding of the Caledonian mountain chains on to the Baltic Shield and to the south another large island extending from central France across southern Germany to Bohemia (see Map 2). To the south of this island southern Europe was submerged, but the detailed picture is not clear.

The story of Upper Palaeozoic times consists, in essence, of two themes – the development and finally

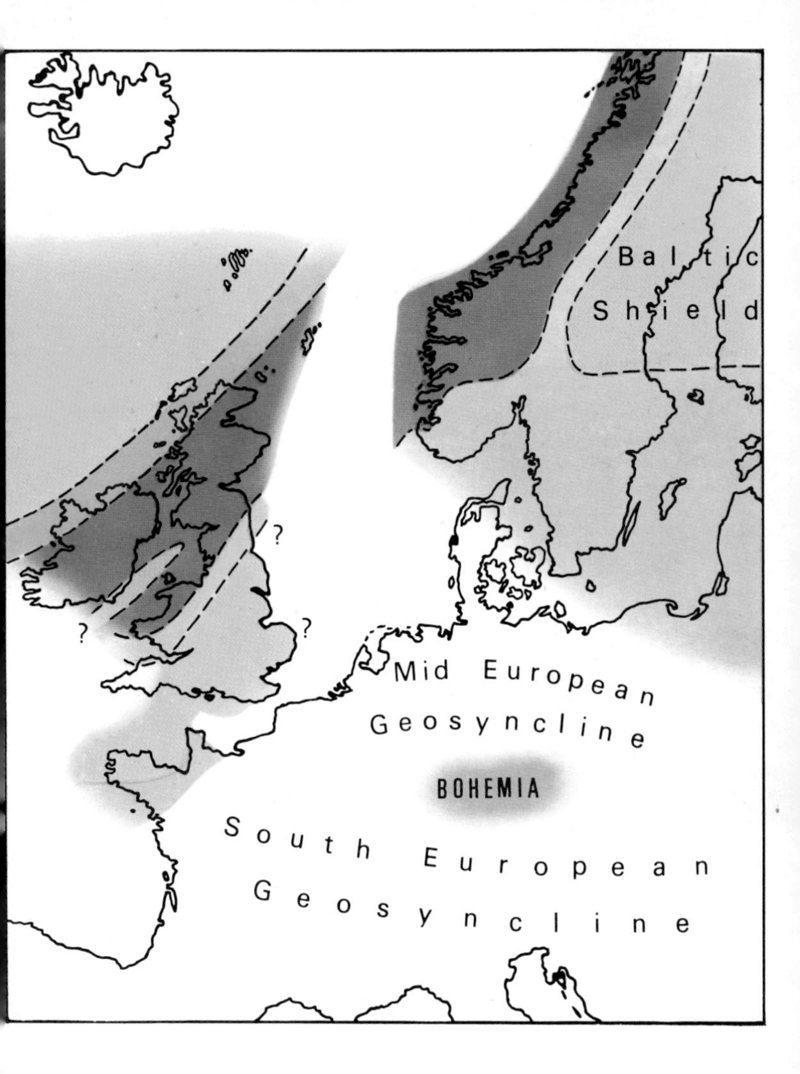

Map 1 Europe in Lower Palaeozoic Times

Land Areas

Shallow Seas

Geosyncline

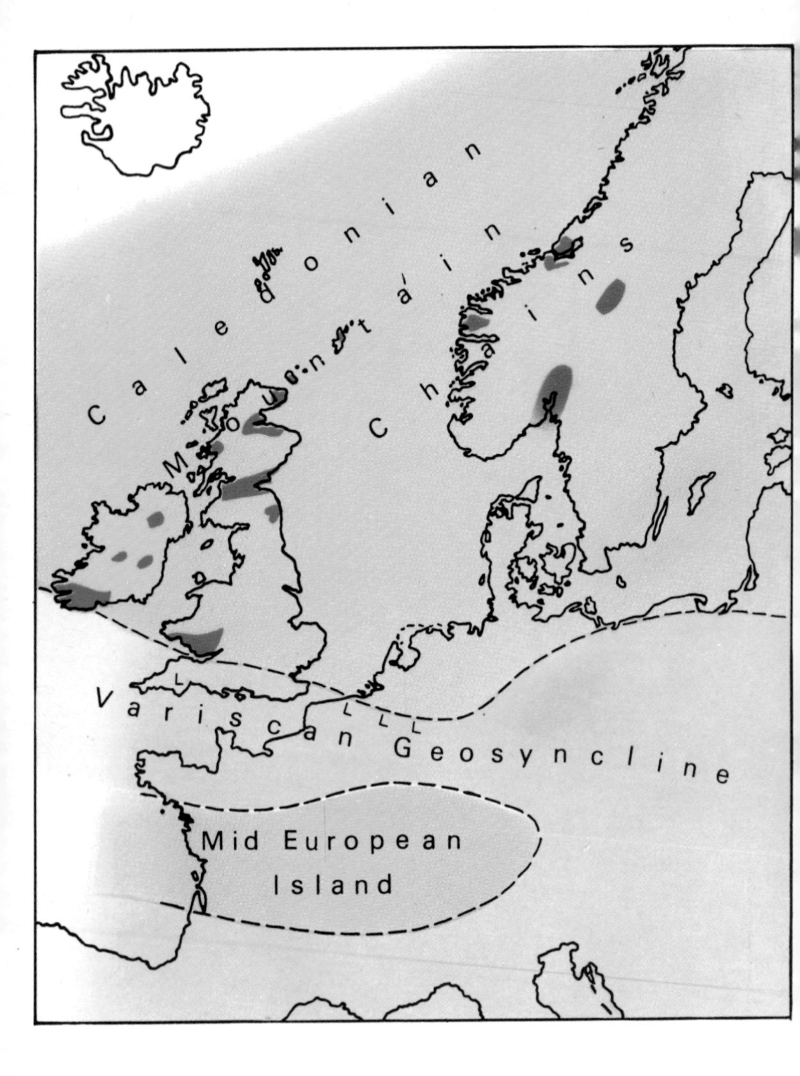

Map 2 Europe during the Devonian Period

Areas of Old Red Sandstone

L Limestone facies

Land Areas

Shallow Seas

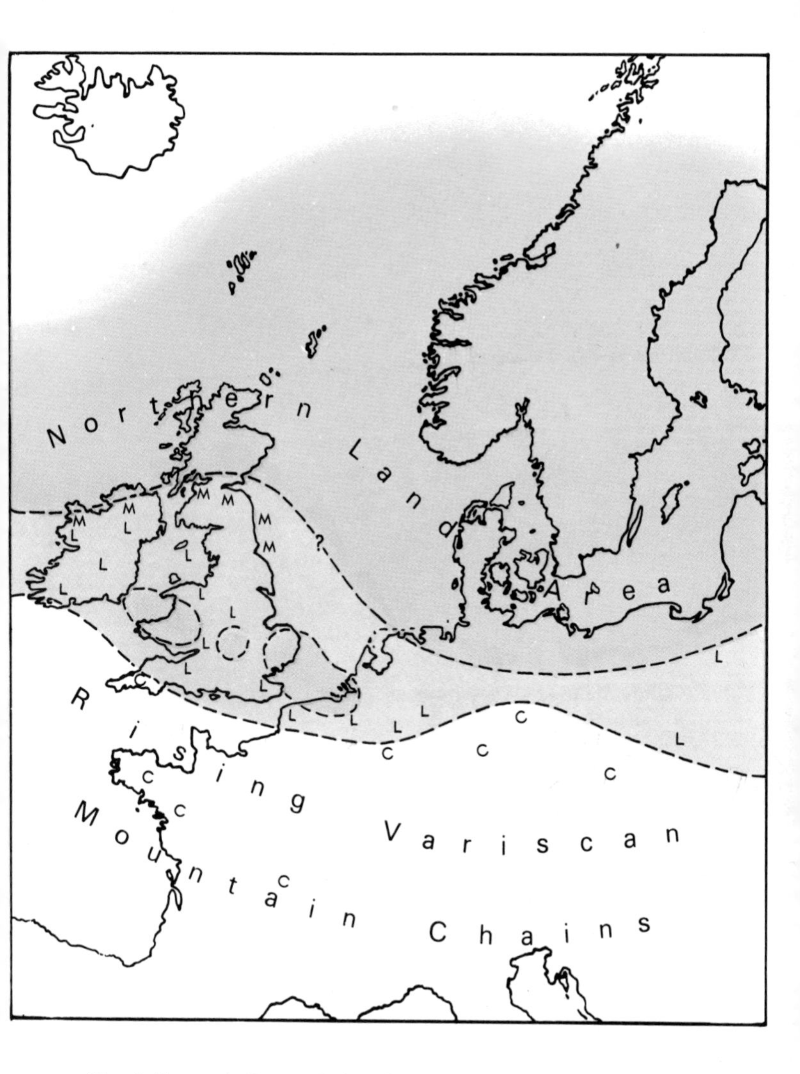

Map 3 Europe in Lower Carboniferous Times

Shallow Seas

Land Areas

C Culm facies

L Limestone facies

M Mixed facies

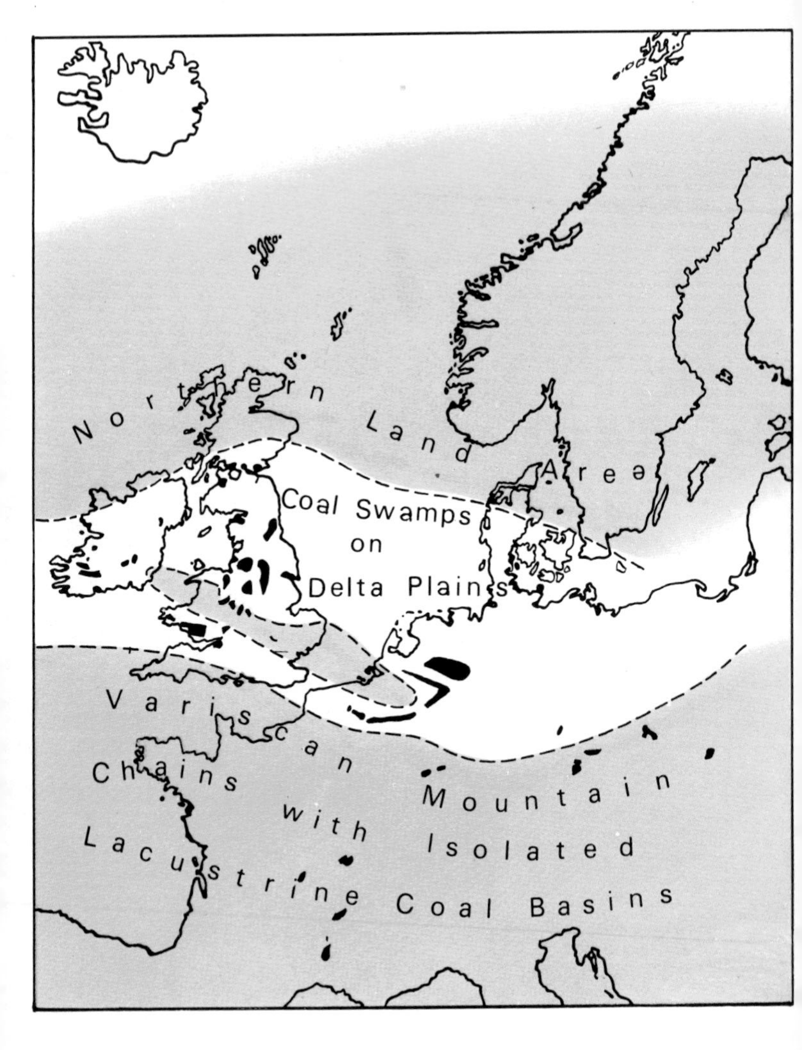

Map 4 Europe in Upper Carboniferous Times

Land Areas

Coal Fields

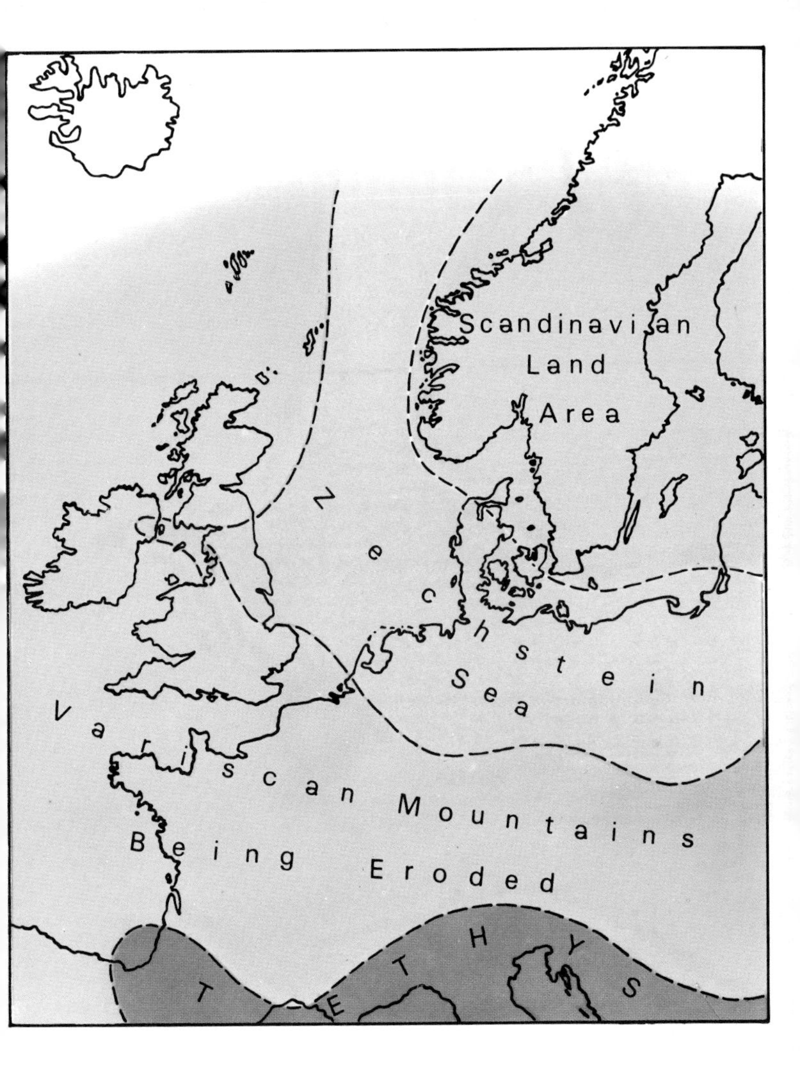

Map 5 Europe during the Permian Period

Land Areas

Shallow Seas

Geosyncline

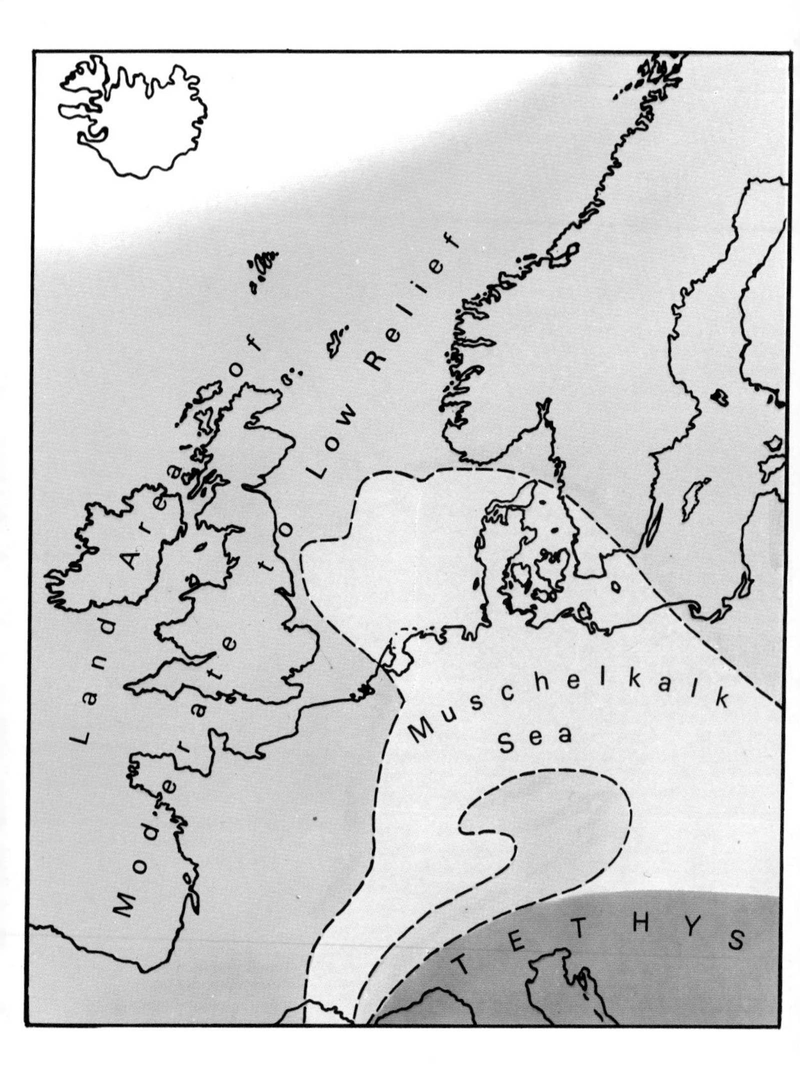

Map 6 Europe during the Triassic Period

Land Areas

Shallow Seas

Geosyncline

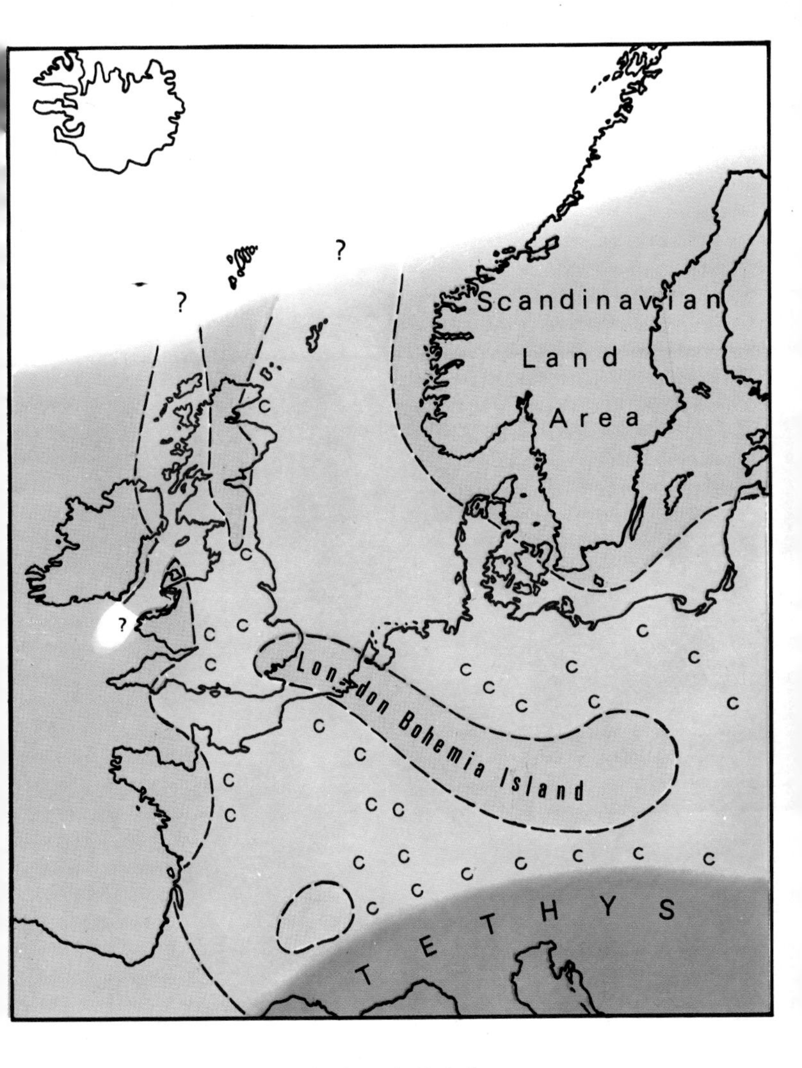

Map 7 Europe during the Jurassic Period

Land Areas

Shallow Seas

Geosyncline

C Coral reefs

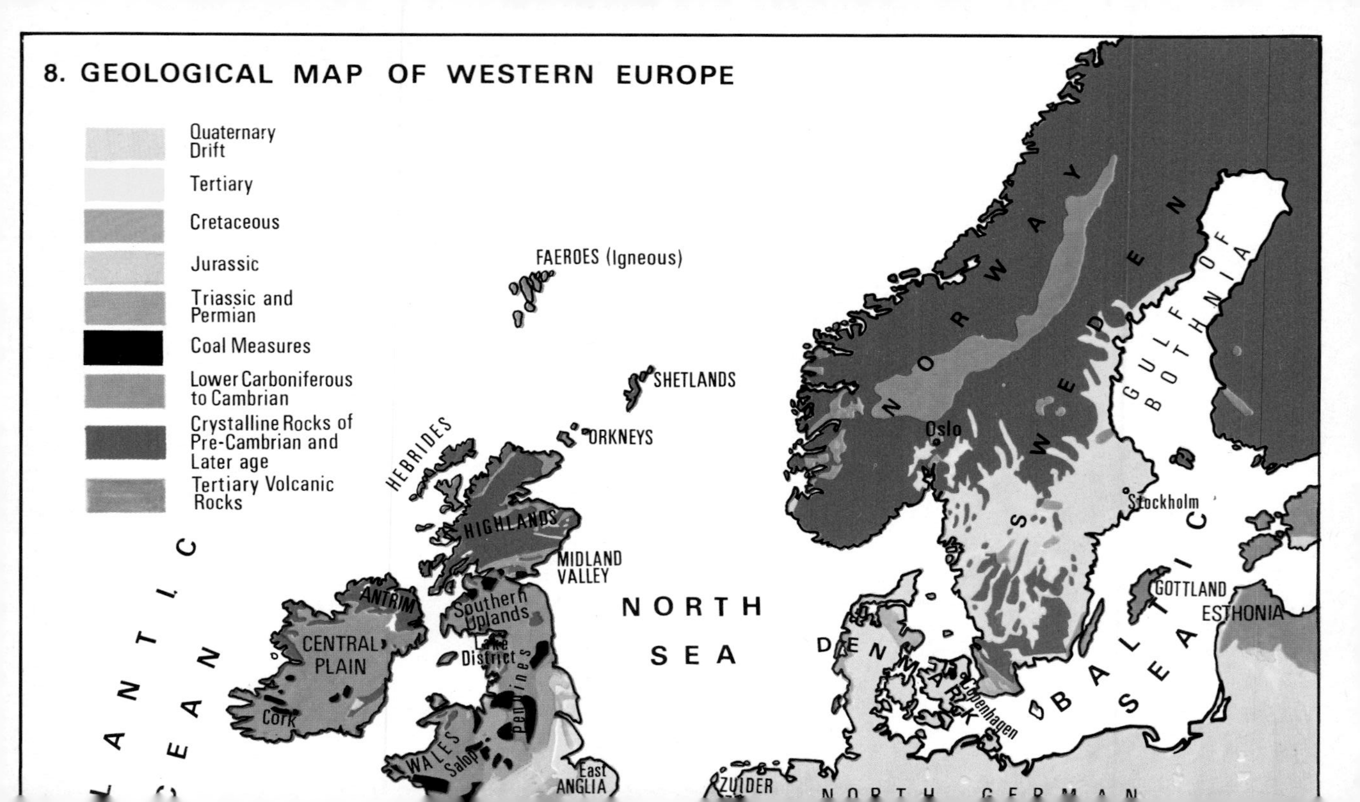
8. GEOLOGICAL MAP OF WESTERN EUROPE
Quaternary Drift
Tertiary
Cretaceous
Jurassic
Triassic and Permian
Coal Measures
Lower Carboniferous to Cambrian
Crystalline Rocks of Pre-Cambrian and Later age
Tertiary Volcanic Rocks
FAEROES (Igneous)
SHETLANDS
ORKNEYS
HEBRIDES
HIGHLANDS
MIDLAND VALLEY
Southern Uplands
Lake District
Pennines
WALES
Salop
East ANGLIA
ANTRIM
CENTRAL PLAIN
Cork
NORTH SEA
NORWAY
SWEDEN
GULF OF BOTHNIA
Oslo
Stockholm
GOTTLAND
ESTHONIA
BALTIC SEA
DENMARK
Copenhagen
ZUIDER
NORTH GERMAN

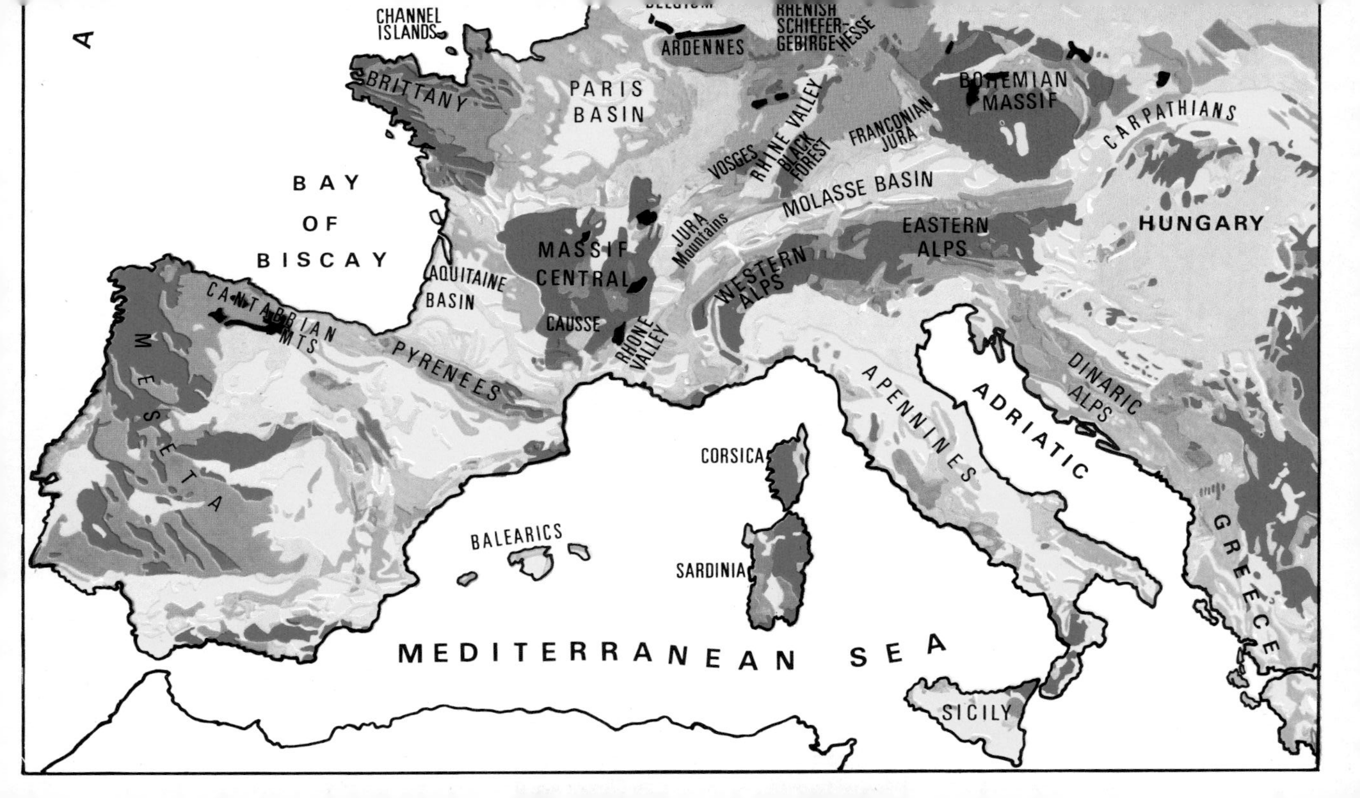
A
CHANNEL ISLANDS
BRITTANY
PARIS BASIN
ARDENNES
RHENISH SCHIEFER-GEBIRGE
HESSE
BOHEMIAN MASSIF
CARPATHIANS
RHINE VALLEY
VOSGES
BLACK FOREST
FRANCONIAN JURA
MOLASSE BASIN
BAY OF BISCAY
JURA Mountains
MASSIF CENTRAL
EASTERN ALPS
HUNGARY
WESTERN ALPS
AQUITAINE BASIN
CAUSSE
RHONE VALLEY
CANTABRIAN MTS
MESETA
PYRENEES
APENNINES
ADRIATIC
DINARIC ALPS
CORSICA
SARDINIA
BALEARICS
GREECE
MEDITERRANEAN SEA
SICILY

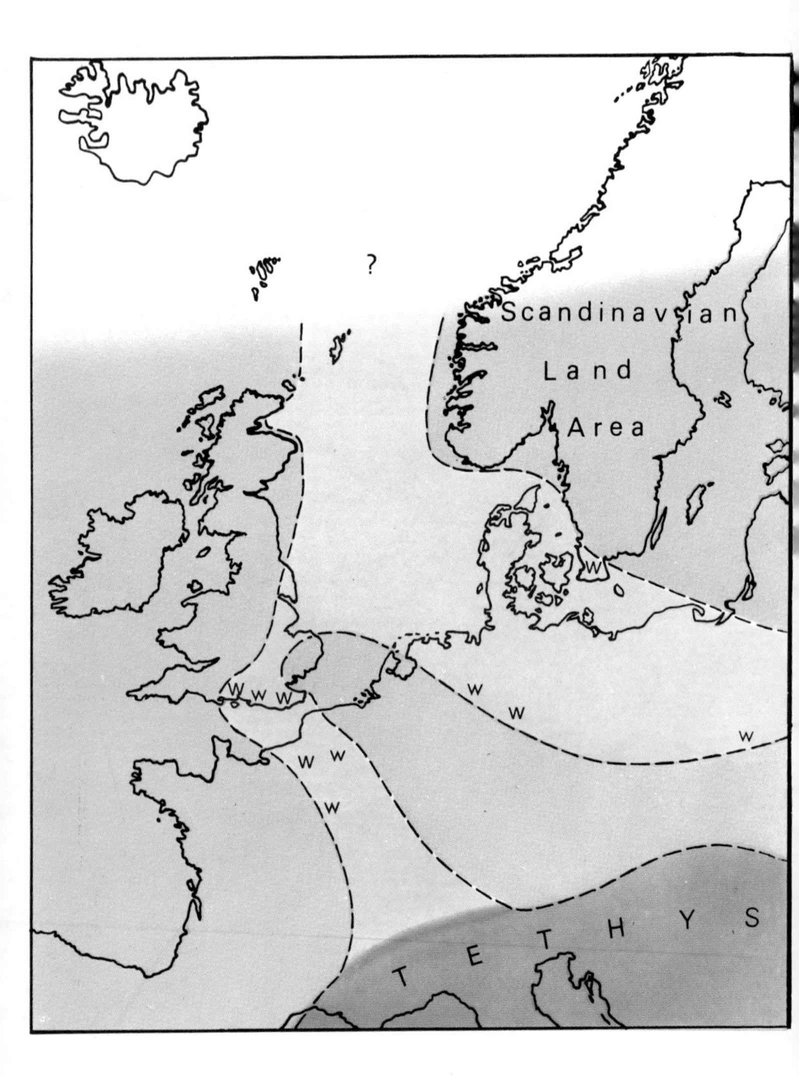

Map 9 Europe during Lower Cretaceous Times

Land Areas

Shallow Seas

Geosyncline

W Wealden beds

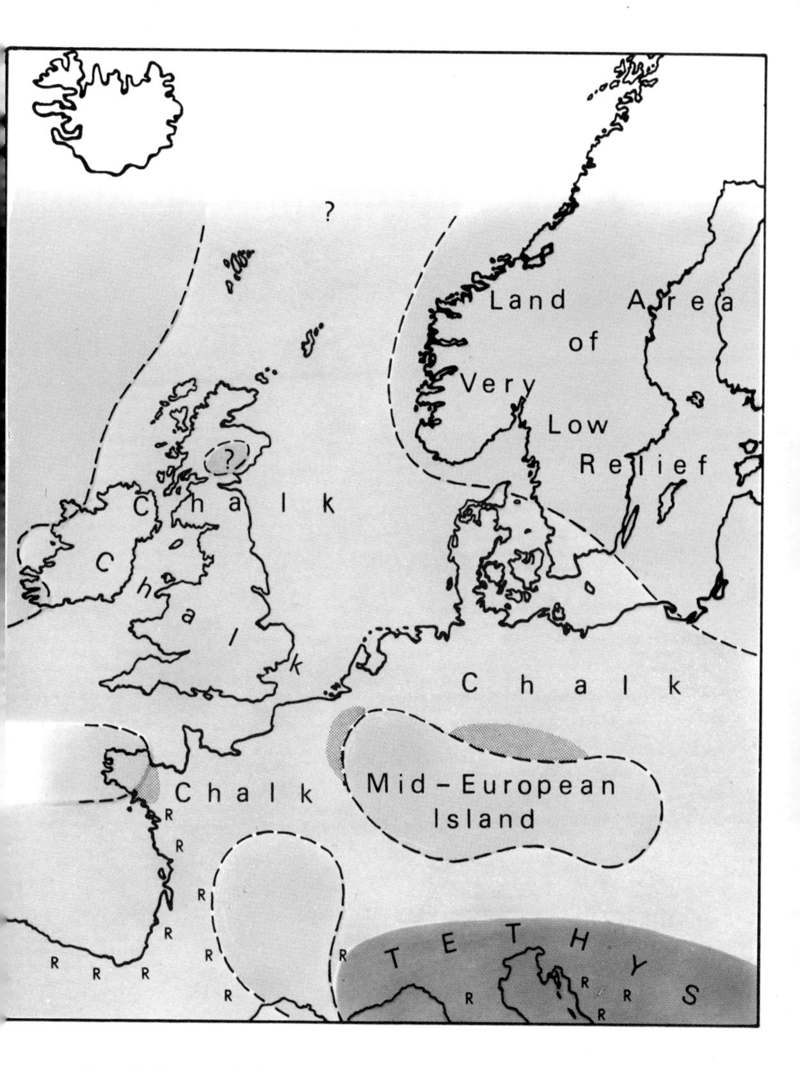

Map 10 Europe during Upper Cretaceous Times

Land Areas

Shallow Seas

Geosyncline

R Rudistid Limestones

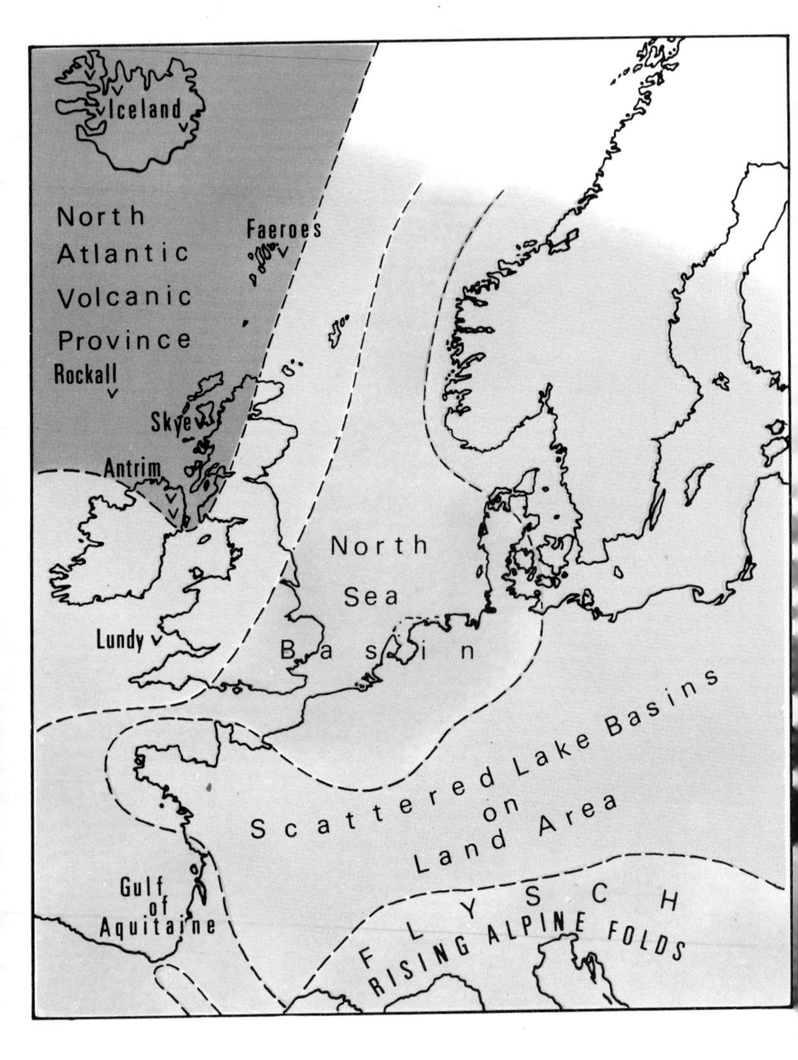

Map 11 Europe during the Lower Tertiary

Land Areas

Shallow Seas

V Centres of igneous activity

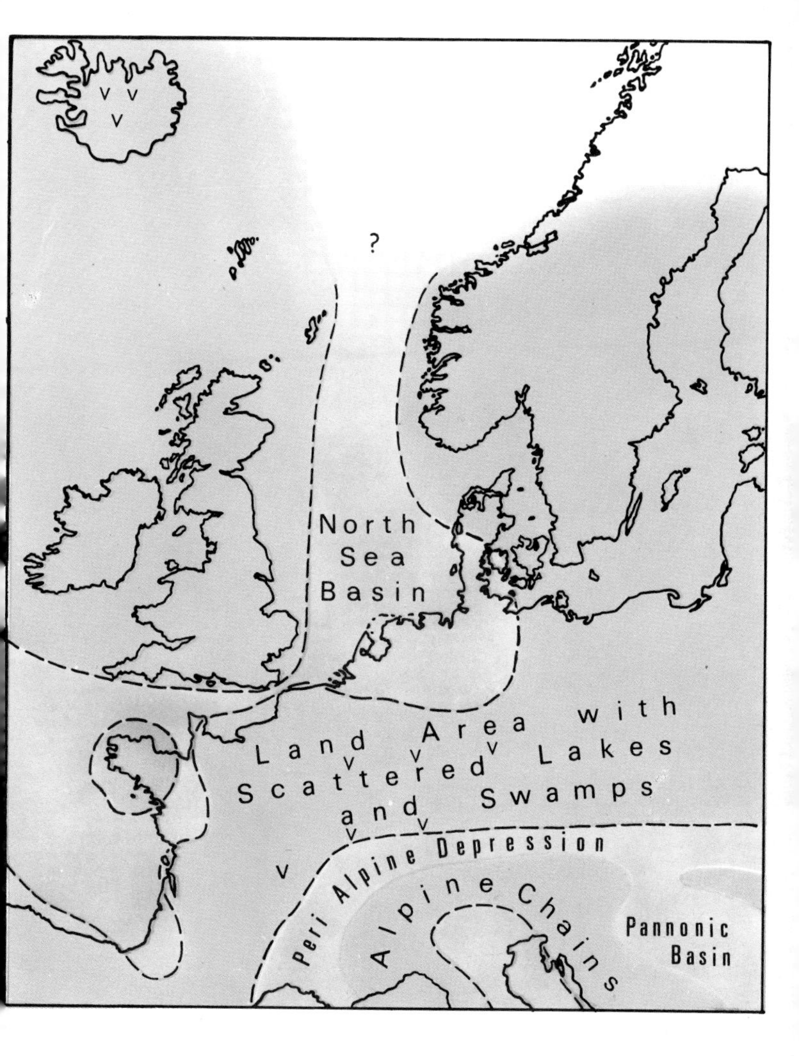

Map 12 Europe during the Upper Tertiary

Land Areas

Shallow Seas

V Centres of igneous activity

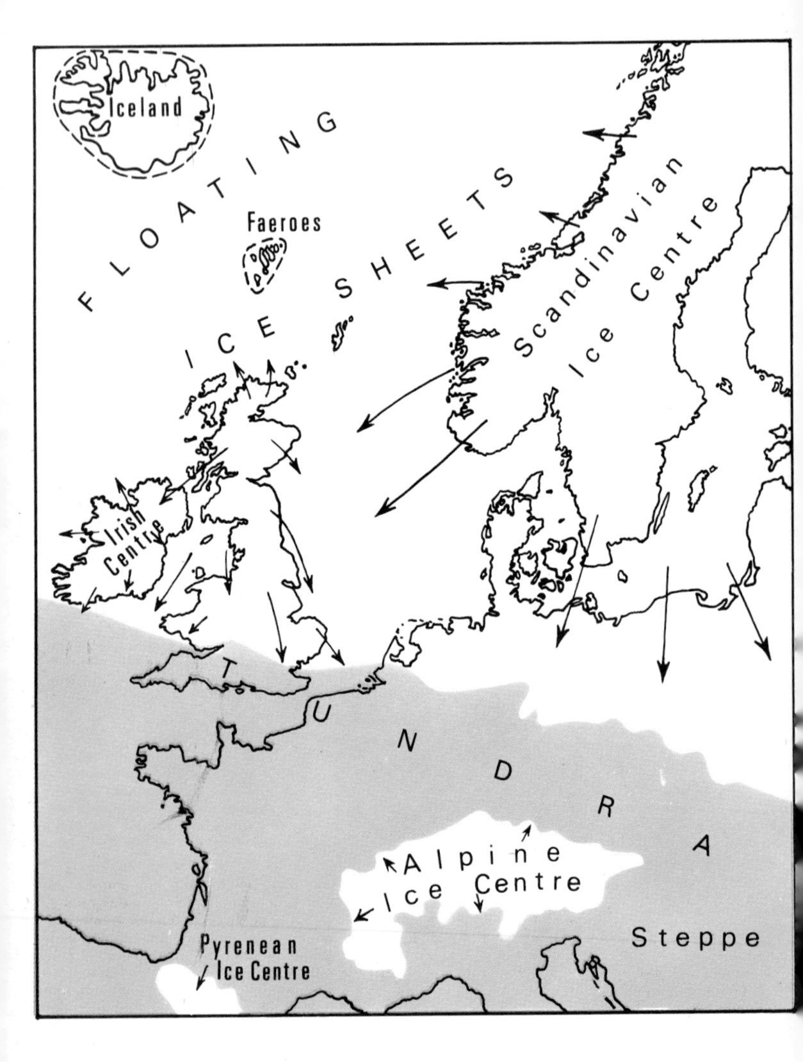

Map 13 Pleistocene Europe

Unglaciated areas

Fossil Forest, Purbeck Beds (Upper Jurassic), Lulworth, Dorset.
(Photo. J. F. K.)

Reptilian Footprints, Purbeck Beds (Upper Jurassic), Worbarrow Bay, Dorset. (Photo. J.F.K.)

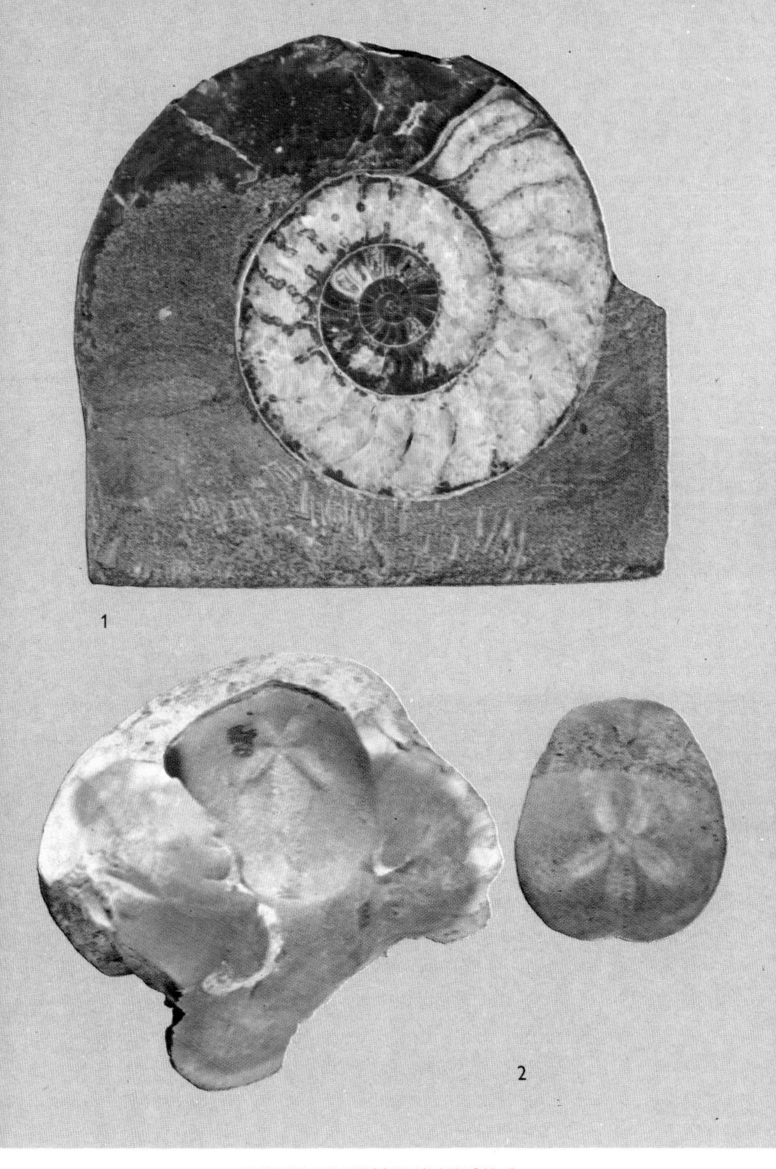

MODES OF FOSSILIZATION, I

1. Cross-section of the ammonite *Asteroceras*, Lias (Lower Jurassic), Lyme Regis, Dorset. $\times \frac{1}{2}$
2. *Micraster* in flint, Chalk (Upper Cretaceous), Hardown, Dorset. $\times 1{:}1$

MODES OF FOSSILIZATION, II

3. 'Roach', Portland Beds (Upper Jurassic), Isle of Portland, Dorset. $\times \frac{1}{2}$
4. Chatwall Sandstone, Upper Ordovician, Soudley, Shropshire. $\times \frac{1}{3}$

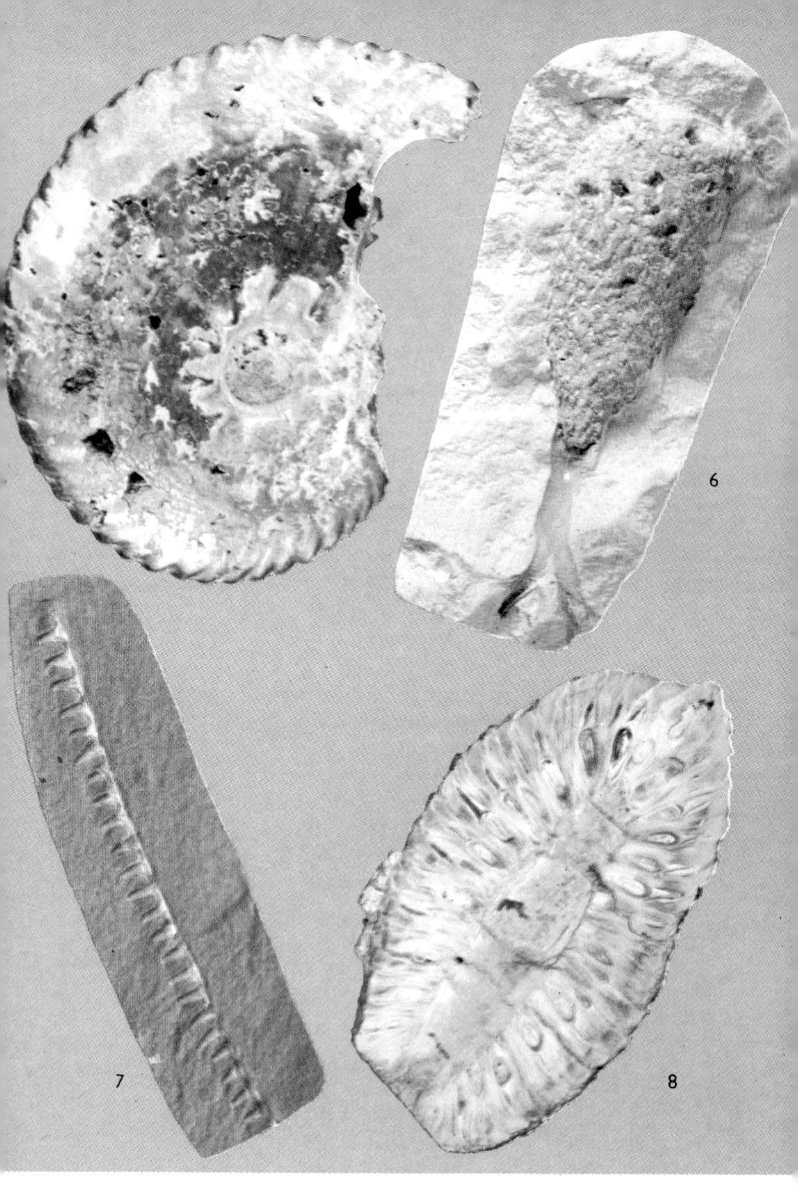

MODES OF FOSSILIZATION, III

5. Pyritized *Anahoplites*, Gault (Lower Cretaceous), Folkestone, Kent. $\times\frac{2}{3}$
6. Ferruginized *Cephalites*, Chalk (Upper Cretaceous), Charlton, Kent. $\times\frac{2}{3}$
7. Pyritized *Monograptus*, Lower Silurian, Pont Erwyd, Cardiganshire, Wales. $\times\frac{2}{3}$
8. Silicified Araucarian Cone, Tertiary, Patagonia. $\times 2$

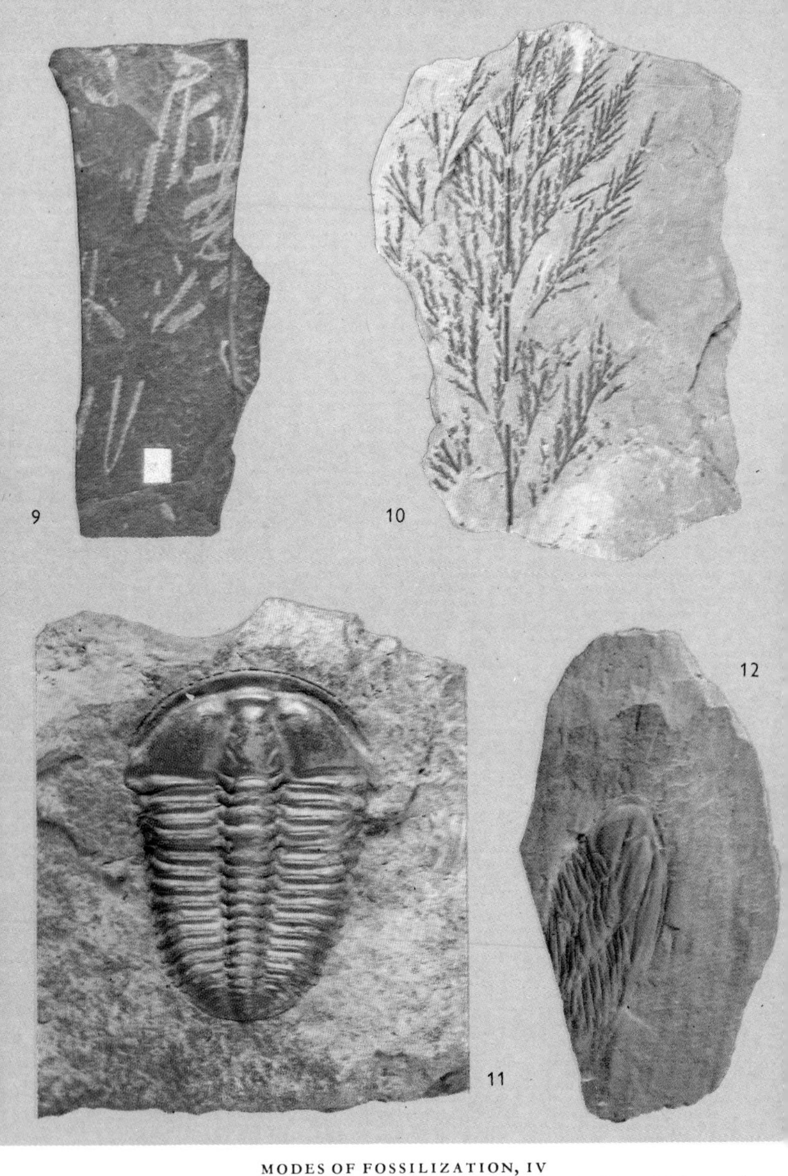

MODES OF FOSSILIZATION, IV

9. *Didymograptus*, Lower Ordovician, Abereiddy, Pembrokeshire, Wales. $\times\frac{2}{3}$
10. *Onychiopsis*, Fairlight Clay (Lower Cretaceous), Hastings, Sussex. $\times\frac{1}{2}$
11. Undistorted trilobite, *Conocoryphe*, Upper Cambrian, Ginetz, Czechoslovakia. $\times\frac{3}{4}$
12. Trilobite distorted by cleavage.
Angelina, Tremadocian (Upper Cambrian), Portmadoc, Caernarvonshire, Wales. $\times 1$

13. Bedded Rocks, Hunstanton, Norfolk. (Photo J. F. K.)

14. Unconformity, Portishead, Somerset. (Photo J. F. K)

DEATH AND LIFE ASSEMBLAGES

15. Bone Bed, Weald Clay (Lower Cretaceous), Henfield, Sussex. $\times \frac{3}{4}$
16. Colony of rhynchonellids, Marlstone (Lower Jurassic), Hook Norton, Oxfordshire. $\times \frac{1}{2}$

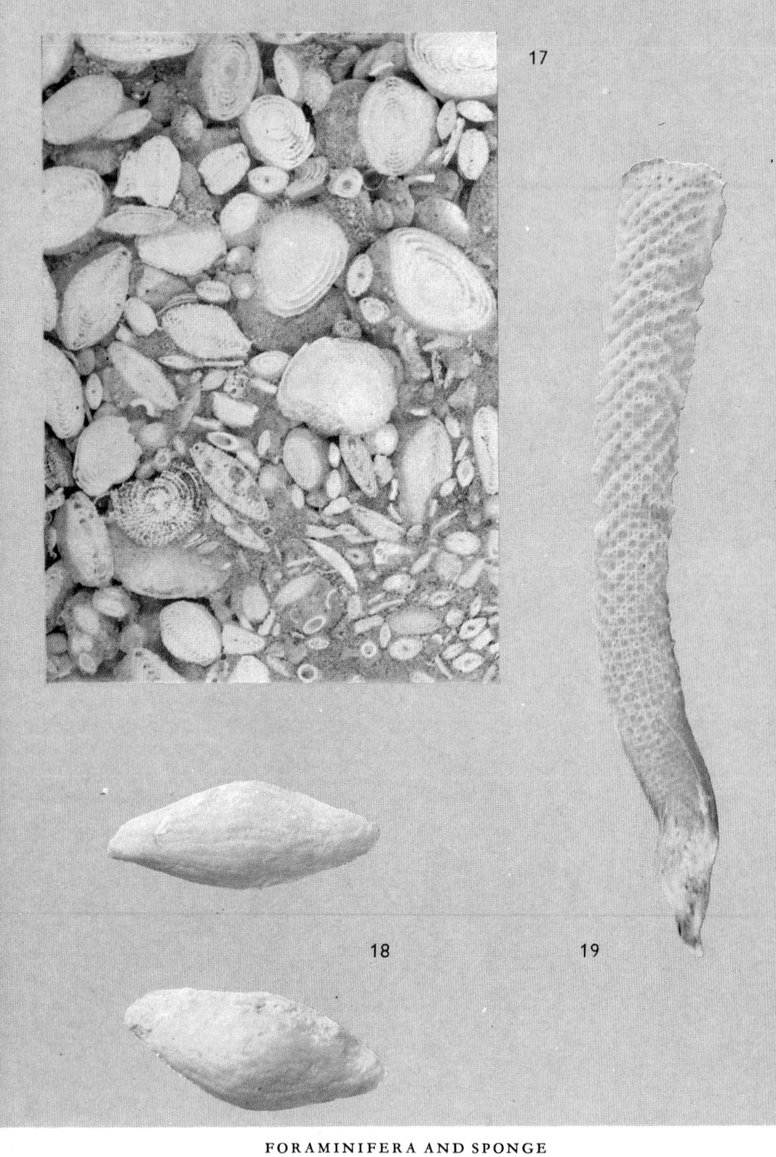

FORAMINIFERA AND SPONGE

17. *Nummulites,* Lutetian (Middle Eocene), Noyon, Somme, France. $\times 1\frac{1}{2}$
18. *Fusulinids,* Upper Carboniferous, locality not known. $\times 3$
19. *Euplectella,* a Recent 'glass sponge', Philippine Islands. $\times \frac{1}{3}$

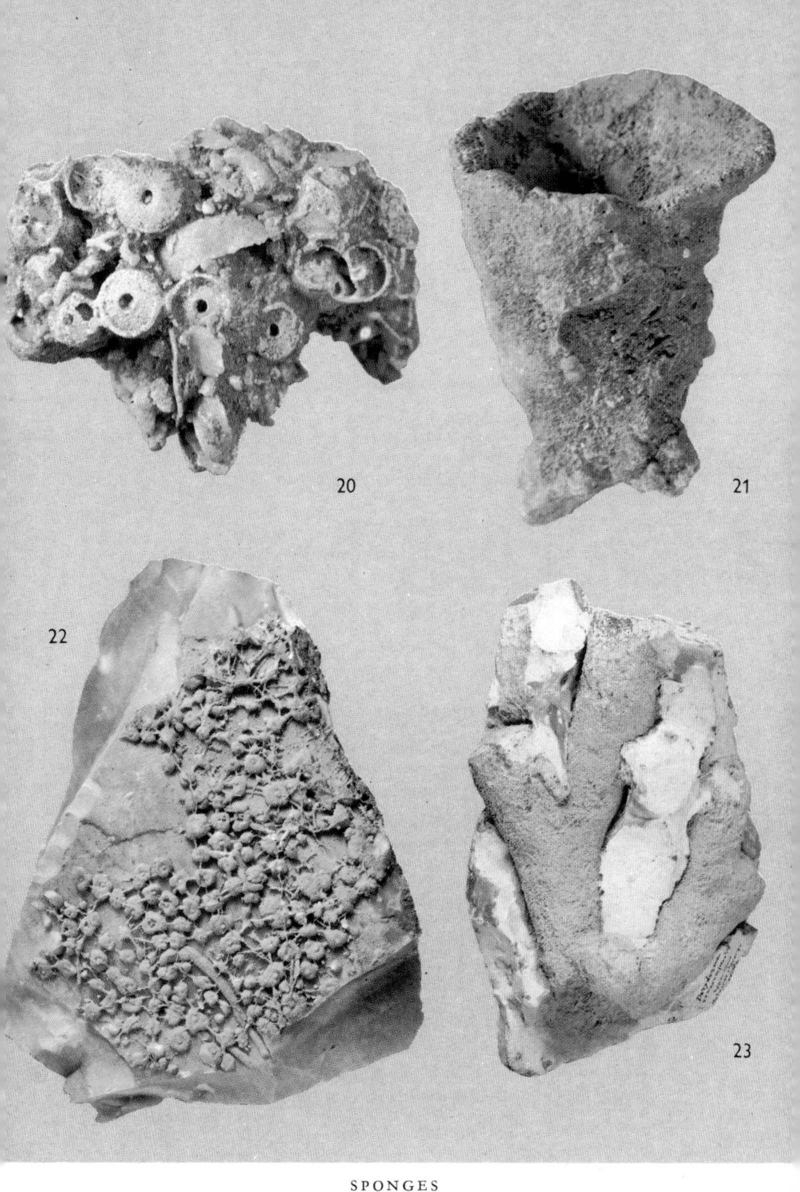

SPONGES

20. *Tremacystia* (*Barroisia*), Sponge Gravels (Lower Cretaceous), Faringdon, Berkshire. $\times 2$
21. *Raphidonema*, Sponge Gravels (Lower Cretaceous), Faringdon, Berkshire. $\times 1$
22. Cast in flint of the borings of *Entobia Cliona*, Chalk (Upper Cretaceous), Brighton, Sussex. $\times 1$
23. *Dorderyma*, Chalk (Upper Cretaceous), Marlborough, Wiltshire. $\times \frac{1}{2}$

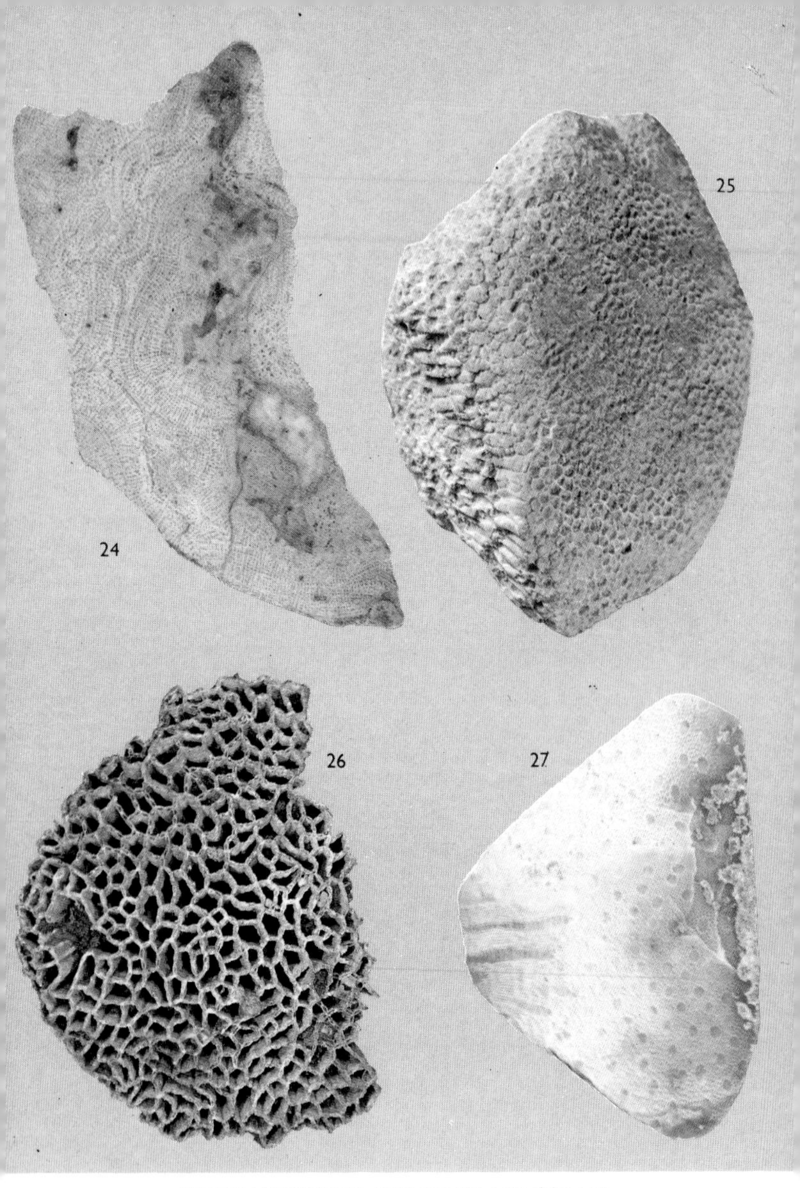

STROMATOPOROID AND TABULATE CORALS

24. *Stromatopora*, Middle Devonian, Bishopsteignton, Devon. $\times 1$
25. *Favosites*, Wenlock Limestone (Middle Silurian), Woolhope, Herefordshire. $\times 1$
26. *Halysites*, Wenlock Limestone (Middle Silurian), Dudley, Warwickshire. $\times 1$
27. *Heliolites*, Middle Devonian, Torquay, Devonshire. $\times 1\frac{1}{3}$

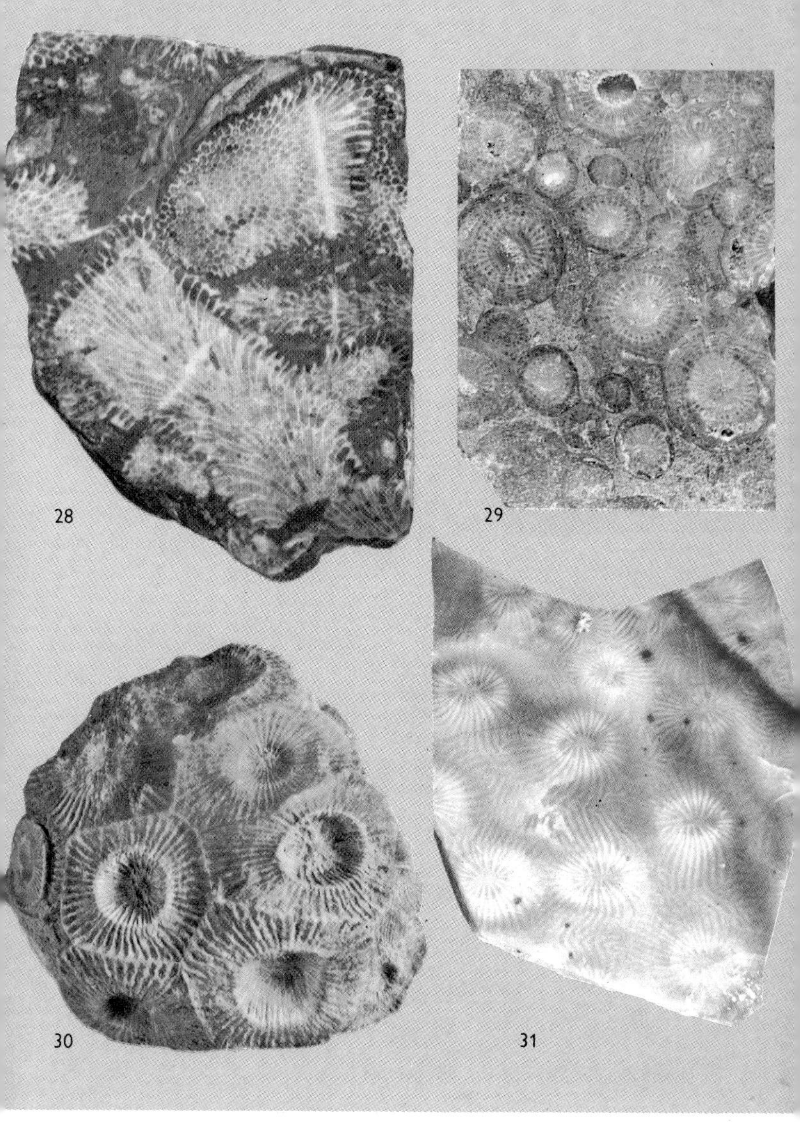

COMPOUND PALAEOZOIC CORALS

28. *Pachypora*, Middle Devonian, Torquay, Devonshire. $\times 1$
29. *Thysanophyllum*, Lower Carboniferous, Shap, Westmorland. $\times \frac{2}{3}$
30. *Acervularia*, Wenlock Limestone (Middle Silurian), Much Wenlock, Shropshire $\times 1\frac{1}{2}$
31. *Phillipsastraea*, Middle Devonian, Torquay, Devon. $\times 3$

SOLITARY CORALS

32. *Palaeosmilia,* Lower Carboniferous, locality not known. $\times \frac{1}{2}$
33. *Zaphrentis* (silicified), Lower Carboniferous, locality not known. $\times 3$
34. *Aulophyllum,* Lower Carboniferous, Bathgate, West Lothian, Scotland. $\times 1\frac{1}{2}$
35. *Dibunophyllum,* Lower Carboniferous, Bathgate, West Lothian, Scotland. $\times 1\frac{1}{2}$

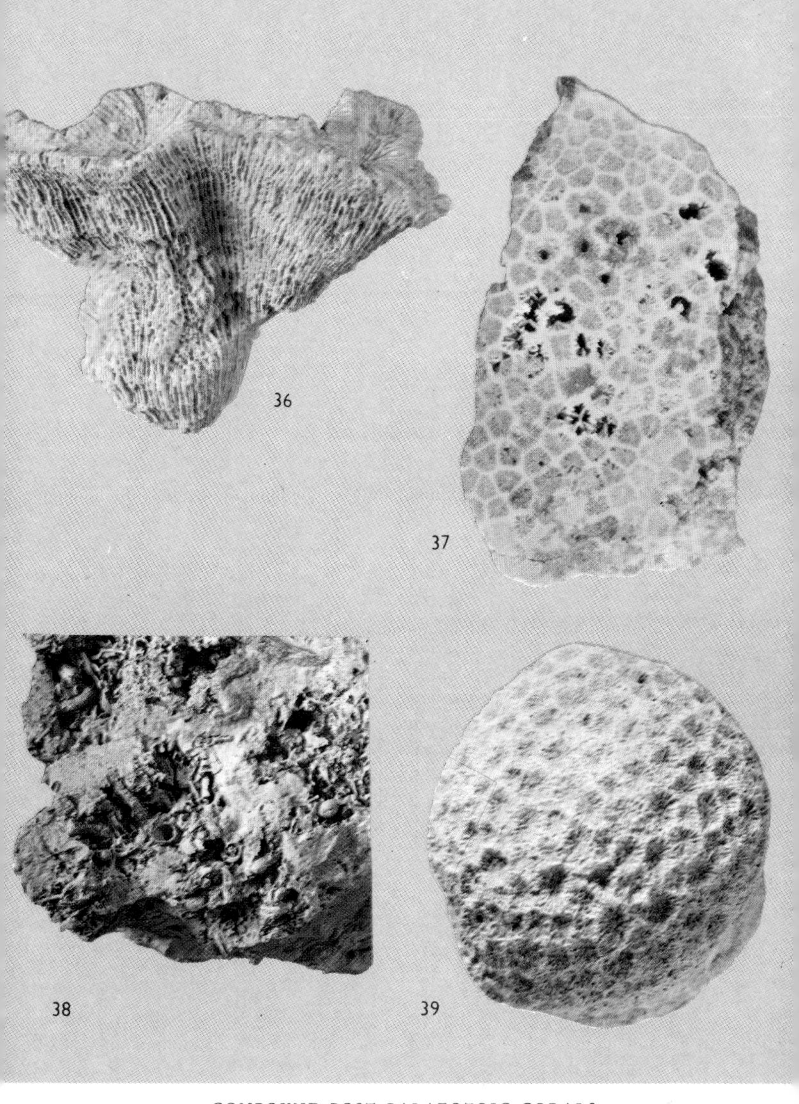

COMPOUND POST-PALAEOZOIC CORALS

36. *Thecosmilia*, Corallian (Upper Jurassic), Steeple Ashton, Wiltshire. $\times \frac{1}{2}$
37. *Isastraea*, Kimmeridgian (Upper Jurassic), Helmsdale, Sutherlandshire, Scotland $\times \frac{3}{4}$
38. *Dendrophyllia*, Danian Chalk, Fakse, Denmark. $\times 1\frac{1}{3}$
39. *Litharea*, Bracklesham Beds (Middle Eocene), Selsea, Sussex. $\times 1\frac{1}{2}$

COMPOUND AND SOLITARY CORALS

40. *Lonsdaleia*, Lower Carboniferous, locality not known. $\times \frac{1}{2}$

41. *Calceola sandalina*, Middle Devonian, locality not known. $\times 1\frac{1}{2}$

42. *Cyclocyathus*, Gault (Lower Cretaceous), Folkestone, Kent. $\times 3$

POLYZOANS

43. Stick polyzoans and brachiopods.
Wenlock Limestone (Middle Silurian), Much Wenlock, Shropshire. $\times 1\frac{1}{2}$
44. *Fenestella*, Lower Carboniferous, Halkyn, Flintshire, Wales. $\times 1\frac{1}{2}$
45. *Fascicularia*, Coralline Crag (Pliocene), Gedgrave, Suffolk. $\times 1$

VERMES

46. *Serpulae,* London Clay (Lower Eocene), Bognor, Sussex. ×3
47. Pipe Rock, Lower Cambrian, Assynt, Sutherland, Scotland. ×1

LOWER PALAEOZOIC BRACHIOPODS

48. *Lingulella*, Upper Cambrian, Portmadoc, Caernarvonshire, Wales. $\times 1\frac{1}{3}$

49. *Dalmanella*, an orthid, Silurian, Thornbury, Gloucestershire. Exterior (left) and interior (right) of dorsal valve. $\times 2$

LOWER PALAEOZOIC BRACHIOPODS

50. Orthids and *Tentaculites,* Upper Ordovician, Soudley, Shropshire. ×1

51. *Camarotoechia nucula,* Upper Silurian, Ludlow, Shropshire. ×1

PENTAMERIDS

52. *Pentamerus oblongus,* Lower Silurian, Plowden, Shropshire. ×2

53. *Conchidium knightii,* Upper Silurian, View Edge, Shropshire. ×2

STROPHOMENIDS

54. Internal and external views of dolomitized *Productus humerosus*, Lower Carboniferous, Breedon, Leicestershire. $\times \frac{2}{3}$
55. *Gigantoproductus*, Lower Carboniferous, Tenby, Pembrokeshire, Wales. $\times \frac{1}{3}$
56. *Chonetes* showing spines, Upper Silurian. Ludlow, Shropshire. $\times 2$

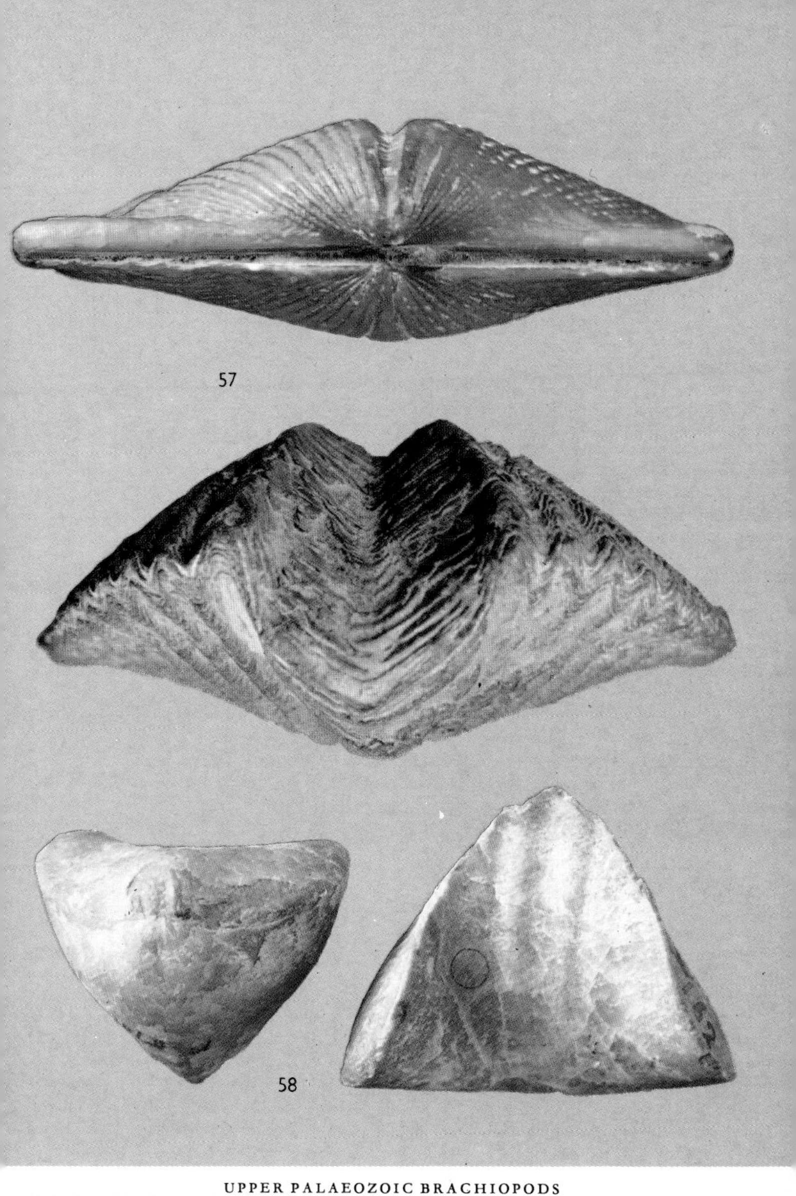

UPPER PALAEOZOIC BRACHIOPODS

57. *Spirifer,* Upper Devonian, Ardennes, Belgium. ×3
58. *Pugnax,* Lower Carboniferous, Clitheroe, Lancashire. ×2

TEREBRATULIDS

59. *Terebratula biplicata,* Red Chalk (Lower Cretaceous), Hunstanton, Norfolk. ×2

60. *Digonella,* Middle Jurassic, Tetbury, Gloucestershire. ×4

MESOZOIC BRACHIOPODS

61. *Sphaeroidothyris,* Middle Jurassic, Bradford Abbas, Dorset. ×3
62. *Rhynchonella,* Lower Chalk (Upper Cretaceous), Beer, Devonshire. ×2
63. *Crania,* Danian Chalk, Fakse, Denmark. ×4

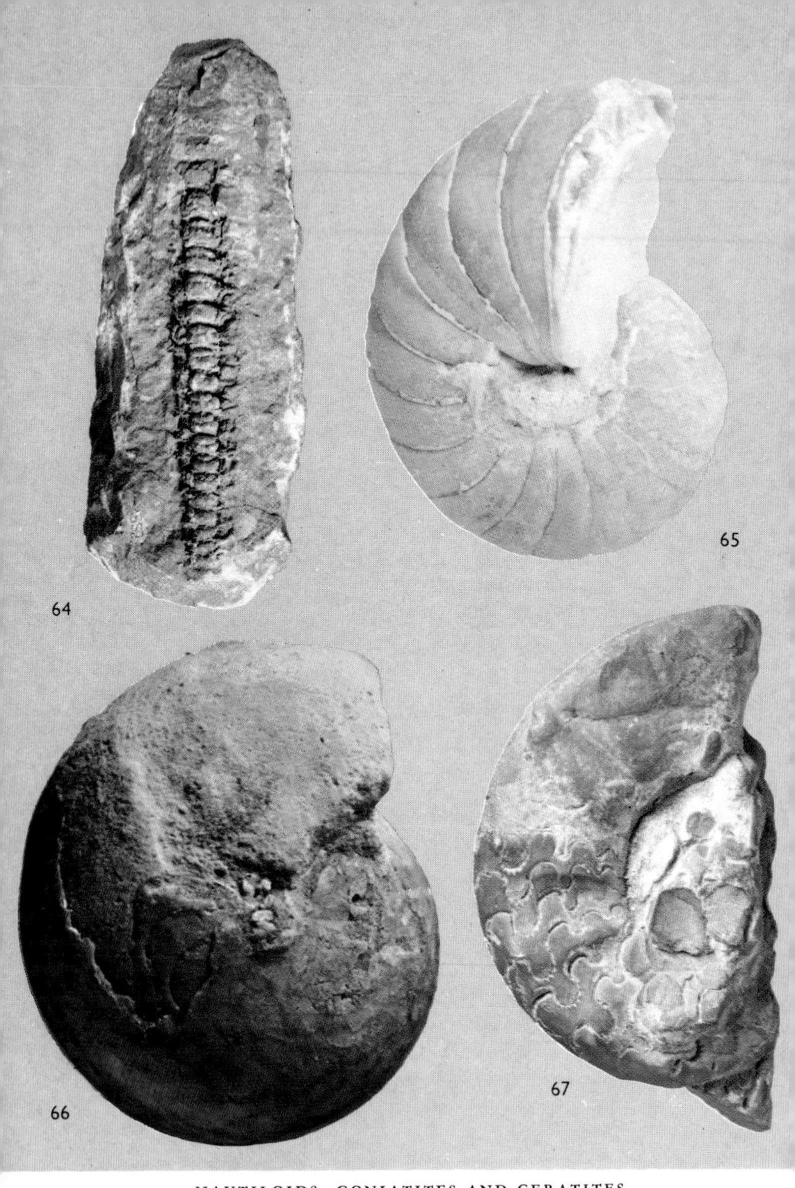

NAUTILOIDS, GONIATITES AND CERATITES

64. *Orthoceras* showing siphuncle, Lower Carboniferous, Kelhead, Annan, Dumfriesshire, Scotland. $\times \frac{2}{3}$

65. *Nautilus*, Inferior Oolite (Middle Jurassic), Cheltenham, Gloucestershire. $\times 1$

66. *Glyphioceras*, a goniatite, Bowland Shales (Upper Carboniferous), Dinkley, Lancashire. $\times 2$

67. *Ceratites*, Muschelkalk (Middle Trias), Olkassen, Saxony, Germany. $\times \frac{2}{3}$

LIASSIC AMMONITES

68. *Promicroceras,* Lower Lias (Lower Jurassic), Lyme Regis, Dorset. ×1
69. *Phylloceras,* Upper Lias (Lower Jurassic), Whitby, Yorkshire. $\times\frac{1}{3}$
70. *Dactylioceras,* Upper Lias, Whitby, Yorkshire. ×1
71. *Harpoceras,* Upper Lias, Whitby, Yorkshire. ×1

UPPER JURASSIC AND CRETACEOUS AMMONITES

72. *Cadoceras*, Kellaways Rock (Upper Jurassic), Scarborough, Yorkshire. $\times 1$
73. *Cosmoceras*, Oxford Clay (Upper Jurassic), Christian Malford, Wiltshire. $\times \frac{3}{4}$
74. *Douvilleiceras*, Gault (Lower Cretaceous), Folkestone Kent. $\times 1$
75. *Euhoplites*, Gault (Lower Cretaceous), Folkestone, Kent. $\times 1\frac{1}{3}$

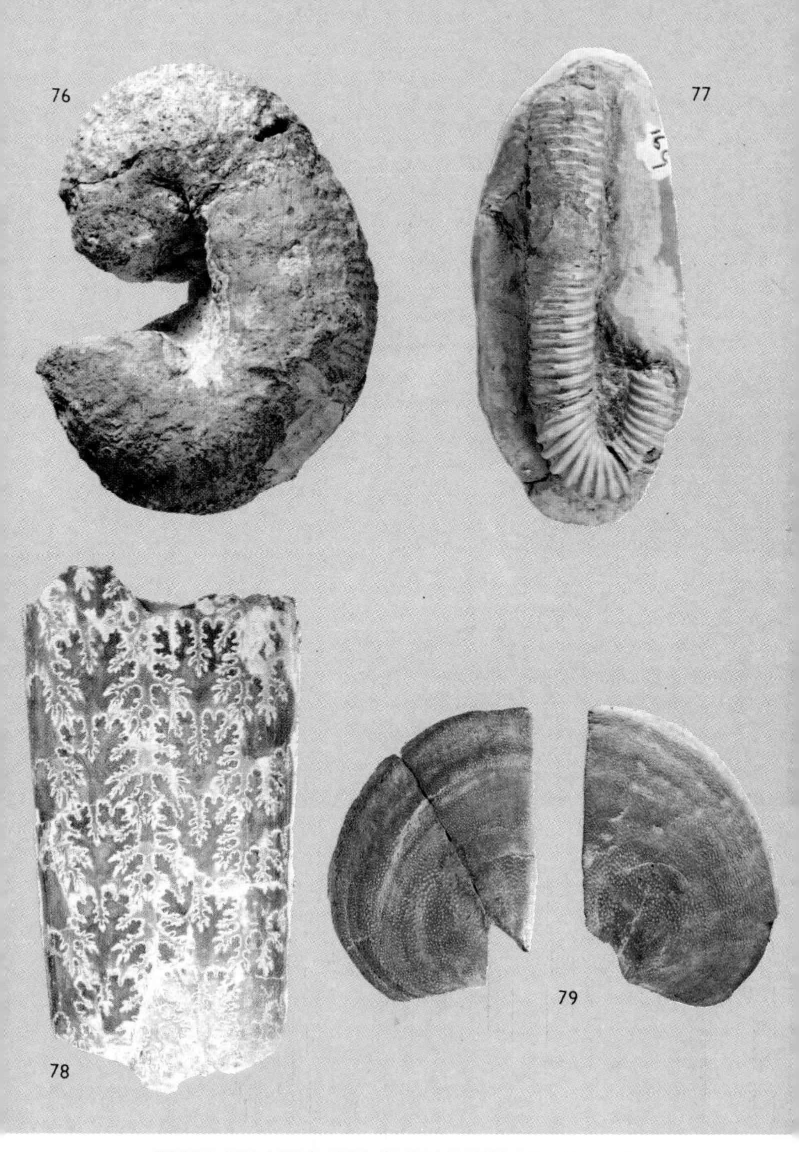

UPPER JURASSIC AND CRETACEOUS AMMONITES

76. *Scaphites*, Chloritic Marl (Upper Cretaceous), Maiden Bradley, Wiltshire. $\times 2$
77. *Hamites*, Gault (Lower Cretaceous), Folkestone, Kent, $\times 1$
78. *Baculites*, Pierre Shale (Upper Cretaceous), Dakota, U.S.A. $\times \frac{2}{3}$
79. Aptychus of an Ammonite, Kimmeridge Clay (Upper Jurassic), Kimmeridge, Dorset. $\times 1$

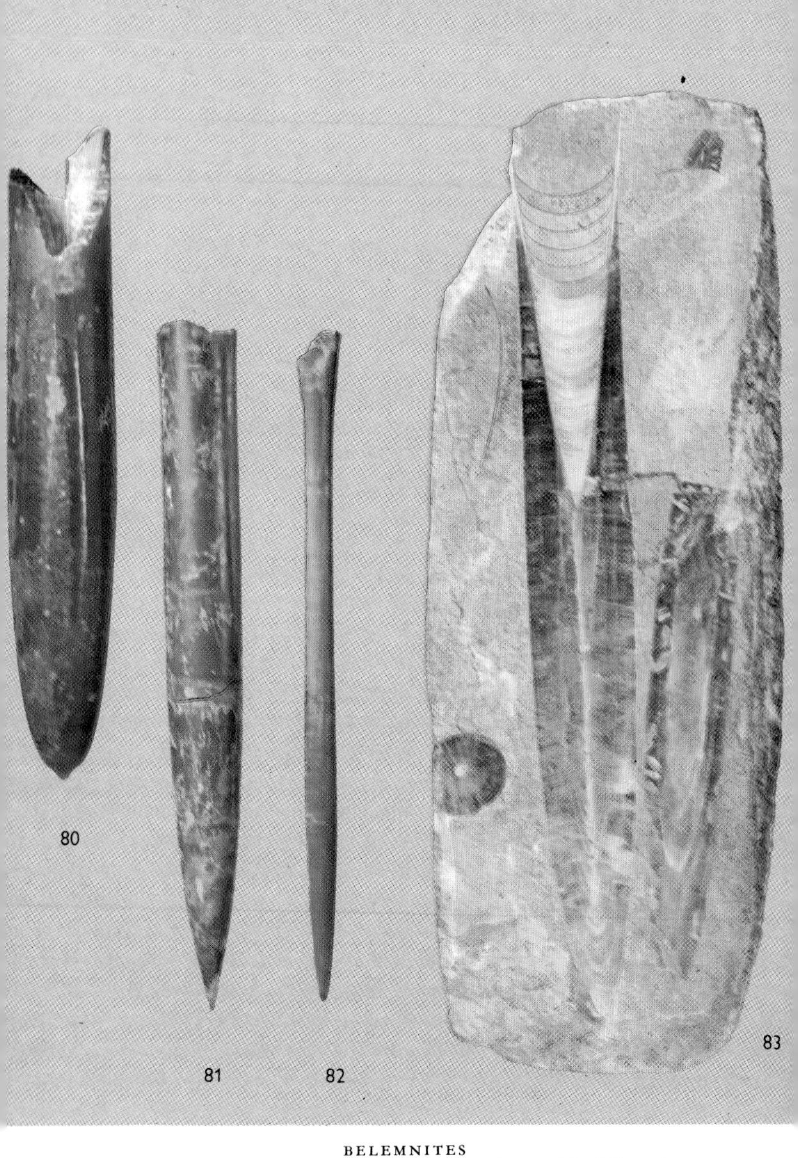

BELEMNITES

80. *Belemnitella*, Upper Chalk (Upper Cretaceous), Thorpe, Norfolk. ×2
81. '*Belemnites*', Hartwell Clay (Upper Jurassic), Aylesbury, Buckinghamshire. ×½
82. *Pseudohastites*, Belemnite Marls (Lower Jurassic), Charmouth, Dorset. ×1
83. Polished cross section of a Belemnite, probably from Corallian Beds (Upper Jurassic), Yorkshire. ×1

ACTIVE LAMELLIBRANCHS

84. *Crassatella*, Barton Beds (Upper Eocene), Barton, Hampshire. $\times 2$

85. *Meretrix*, Auversian (Middle Eocene), Auvers, France. $\times 4$

LAMELLIBRANCHS WITH BYSSAL ATTACHMENT

86. *Entolium* (*Pecten*), Upper Greensand (Lower Cretaceous), Anstey, Wiltshire. $\times 1\frac{1}{2}$

87. *Neithea* (*Pecten*), Upper Greensand (Lower Cretaceous), Lulworth, Dorset. $\times \frac{2}{3}$

88. *Pteria* (*Oxytoma*), Oxford Clay (Upper Jurassic), Woodham, Buckinghamshire. $\times 2$

89. *Musculus* (*Modiola*), Barton Clay (Upper Eocene), Barton, Hampshire. $\times 3$

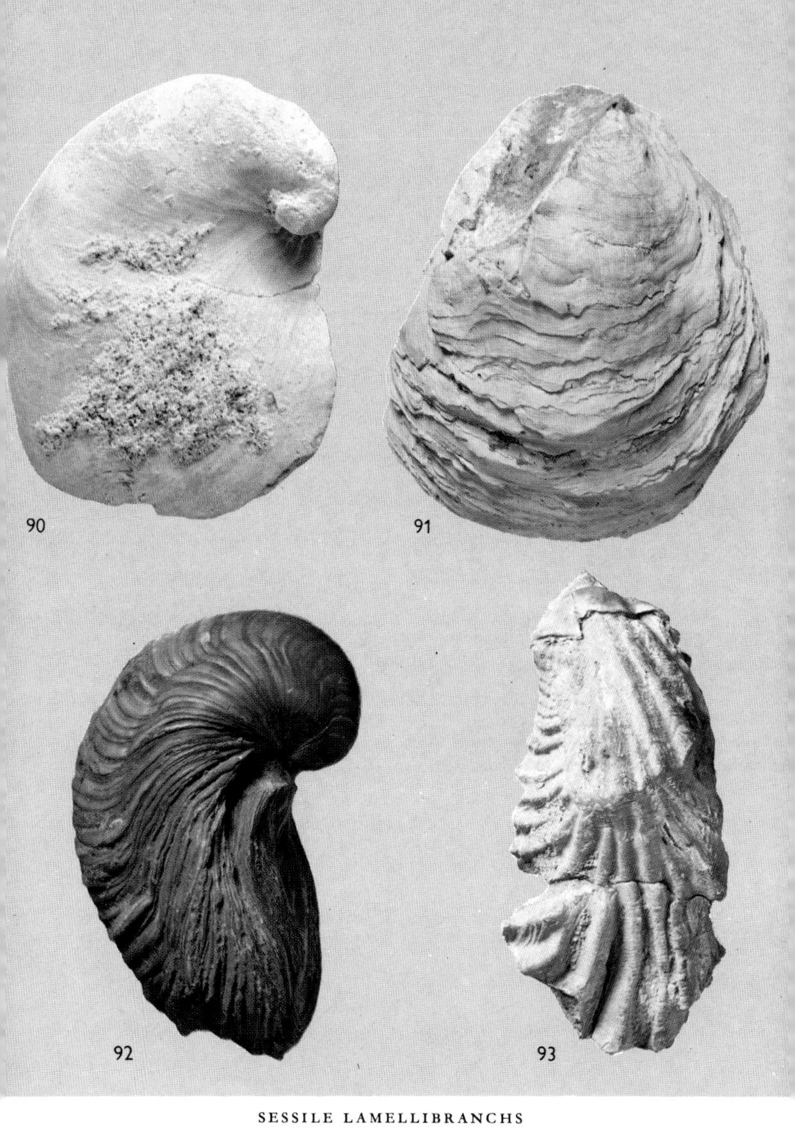

SESSILE LAMELLIBRANCHS

90. *Exogyra*, Upper Greensland (Lower Cretaceous), Beaminster, Dorset. $\times 2$
91 *Ostrea*, Woolwich Beds (Lower Eocene), Charlton, Kent. $\times {}^{2}/_{5}$
92. *Gryphaea*, Lower Lias (Lower Jurassic), Lyme Regis, Dorset. $\times 1$
93. *Lopha*, Inferior Oolite (Middle Jurassic), Cheltenham, Gloucestershire. $\times 1$

BURROWING AND BORING LAMELLIBRANCHS

94. *Pholadomya,* Inferior Oolite (Middle Jurassic), Cheltenham, Gloucestershire. $\times 1\frac{1}{2}$
The 'gape' between the valves is painted red.
95. *Panopea,* London Clay (Lower Eocene), Swanwick. Hampshire. $\times 1$
96. *Teredo* borings in fossil wood. London Clay (Lower Eocene), Sheppey, Kent. $\times 1$

PALAEOZOIC AND MESOZOIC LAMELLIBRANCHS

97. '*Modiolopsis*', Upper Ludlovian (Upper Silurian), probably Carmarthenshire, Wales. $\times \frac{3}{8}$

98. *Dunbarella* (*Pterinopecten*), Millstone Grit (Upper Carboniferous), Ashover, Derbyshire. $\times 1$

99. *Carbonicola*, Coal Measures (Upper Carboniferous), Biddulph, Staffs. $\times {}^{2}/_{7}$

100. *Unio*, Upper Purbeck Beds (Lower Cretaceous), Swanage, Dorset. $\times \frac{1}{3}$

CONTRAST IN DENTITION

101. *Glycimeris,* taxodont dentition, Red Crag (Pleistocene), Walton-on-the-Naze, Essex. $\times 1$

102. *Trigonia,* schizodont dentition, Inferior Oolite (Middle Jurassic), Cotswold Hills. $\times 2$

MESOZOIC LAMELLIBRANCHS

103. Oyster encrusted surface, Middle Jurassic, Snowshill, Gloucestershire. $\times \frac{1}{5}$

104. *Inoceramus*, Chalk (Upper Cretaceous), Seaton, Devon. $\times \frac{1}{2}$

MESOZOIC AND TERTIARY LAMELLIBRANCHS

105. *Nucula*, Internal Mould, Gault (Lower Cretaceous), Folkestone, Kent. ×3
106. *Chama*, Barton Clay (Upper Eocene), Barton, Hampshire. ×½
107. *Astarte,* Coralline Crag (Pliocene), Orford, Suffolk. ×3

TERTIARY LAMELLIBRANCHS

108. *Venericor*, Bracklesham Beds (Middle Eocene), Selsey, Sussex. $\times \frac{3}{4}$
109. *Corbicula*, Blackheath Beds (Lower Eocene), Lessness, Kent. $\times 2$

HOLOSTOMATOUS AND SIPHONOSTOMATOUS GASTROBODS

110. Left *Turritella*, Right *Cerithium*
Bartonian (Upper Eocene), Gisois, Normandy, France $\times 1\frac{1}{4}$

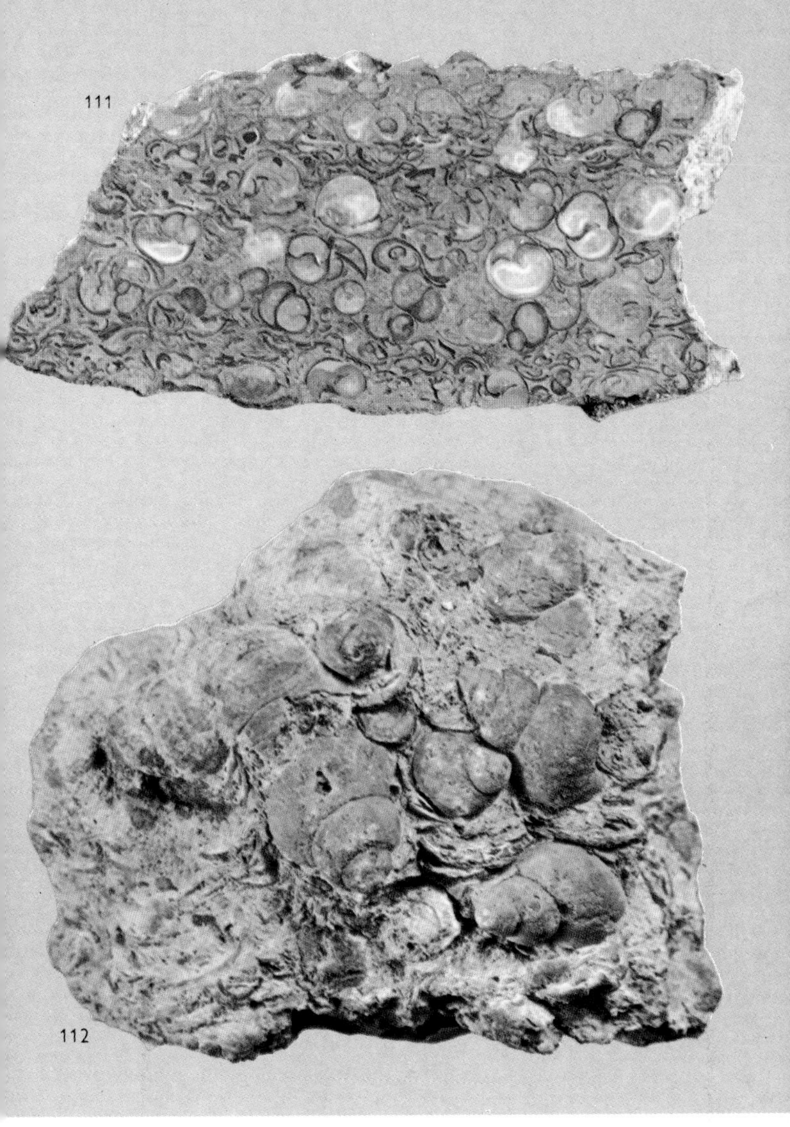

Viviparus IN SUSSEX MARBLE

111. Polished slab. $\times 3$
112. Weathered surface
Weald Clay (Lower Cretaceous), Petworth, Sussex. $\times {}^3/_2$

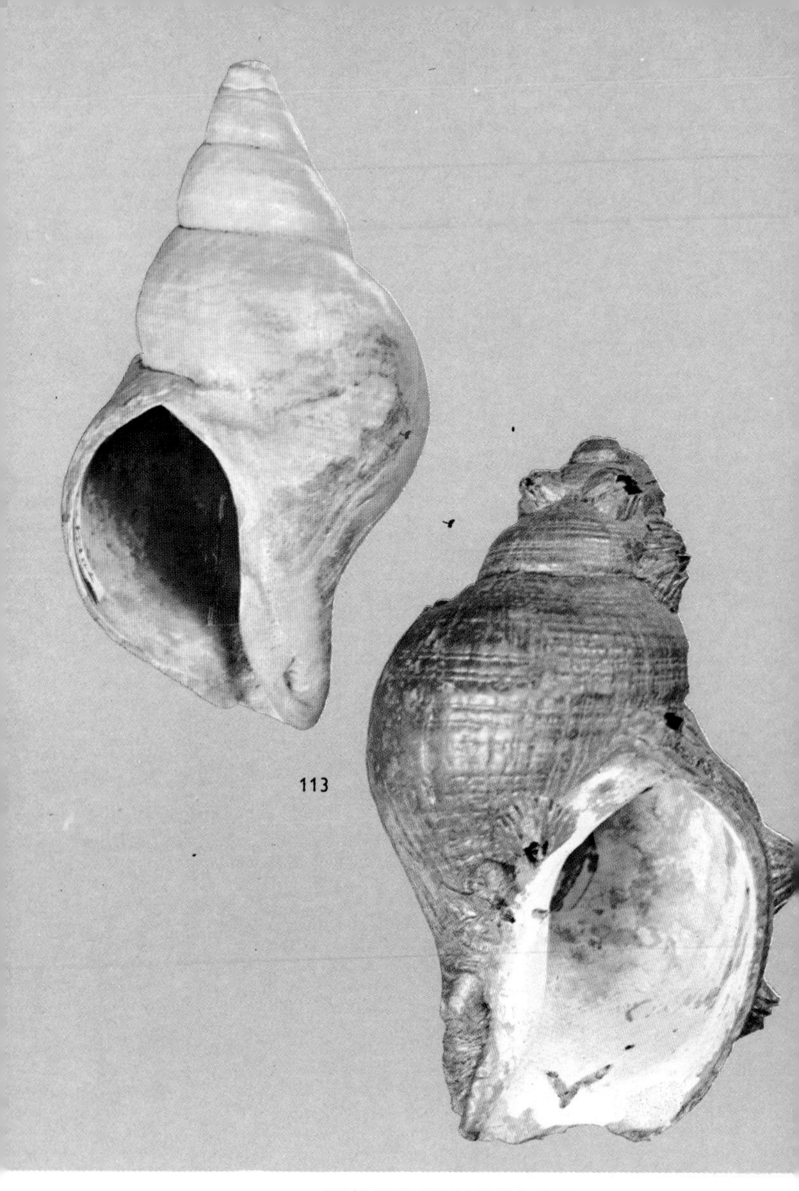

COILING IN GASTROPODS

113. Left *Neptunea contraria*. $\times {}^{3}/_{2}$
Right *Neptunea despecta* encrusted with barnacles.
Red Crag (Pleistocene), Walton-on-the-Naze, Essex.

SCAPHOPOD AND GASTROPODS

114. *Antalis (Dentalium)*, Barton Beds Upper Eocene), Barton, Hampshire. $\times 2$
115. *Pleurotomaria*, Lias (Lower Jurassic), Evesham, Worcestershire. $\times 1$
116. *Emarginula*, Red Crag (Pleistocene), Walton-on-the-Naze, Essex. $\times 7$
117. *Platyschisma*, Downtonian (Devonian), Knighton, Radnorshire, Wales. $\times 1\frac{1}{2}$

TERTIARY GASTROPODS

118. *Aporrhais,* Red Crag (Pleistocene), Walton-on-the-Naze, Essex. ×3
119. *Volutaspina* ×2
120. *Fusinus* ×2
121. *Murex* ×3
Barton Beds (Upper Eocene), Barton, Hampshire.

TERTIARY GASTROPODS

122. *Galba (Lymnaea)*, Oligocene, Bembridge, Isle of Wight. ×1
123. *Potamides*, Headon Beds (Oligocene), Alum Bay, Isle of Wight ×5
124. *Littorina*, Norwich Crag (Pleistocene), Bramerton, Norfolk. ×4

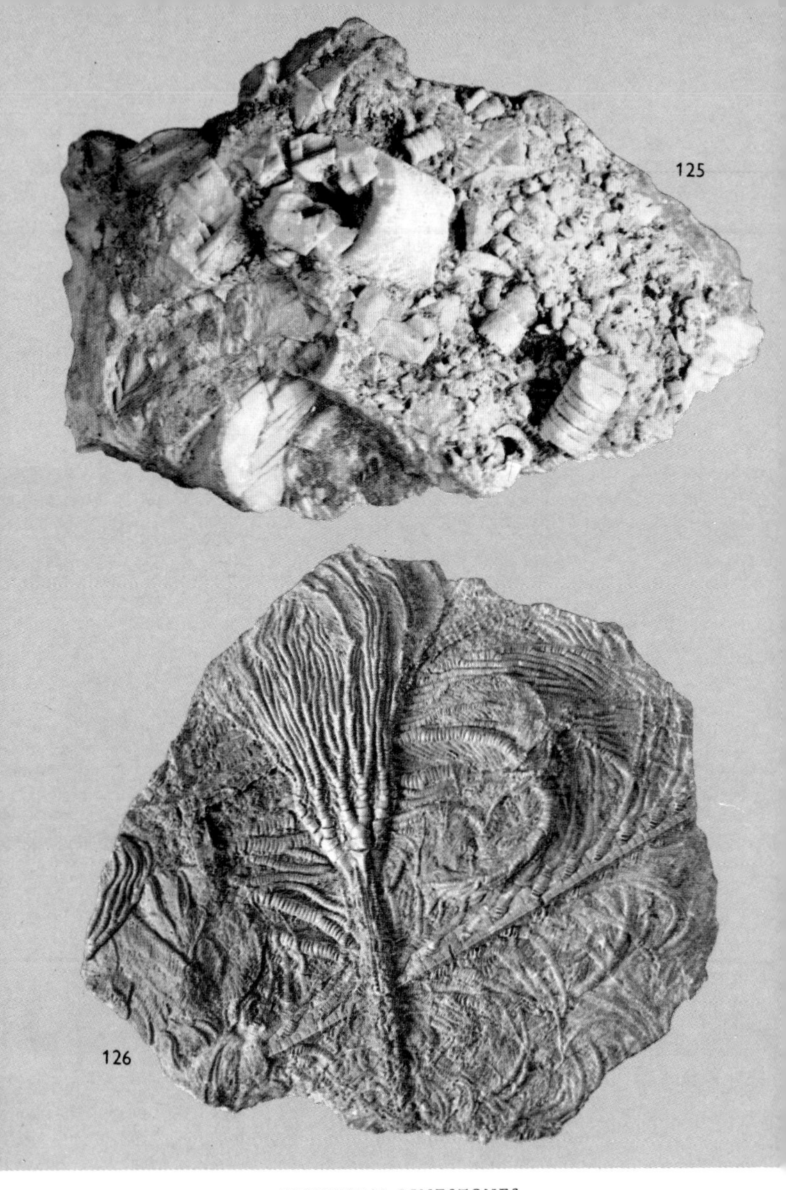

CRINOIDAL LIMESTONES

125. Weathered Carboniferous Limestone, Tenby, Pembrokeshire, Wales. $\times 1\frac{1}{3}$

126. *Pentacrinus* band, Lias (Lower Jurassic), Lyme Regis, Dorset. $\times 1$

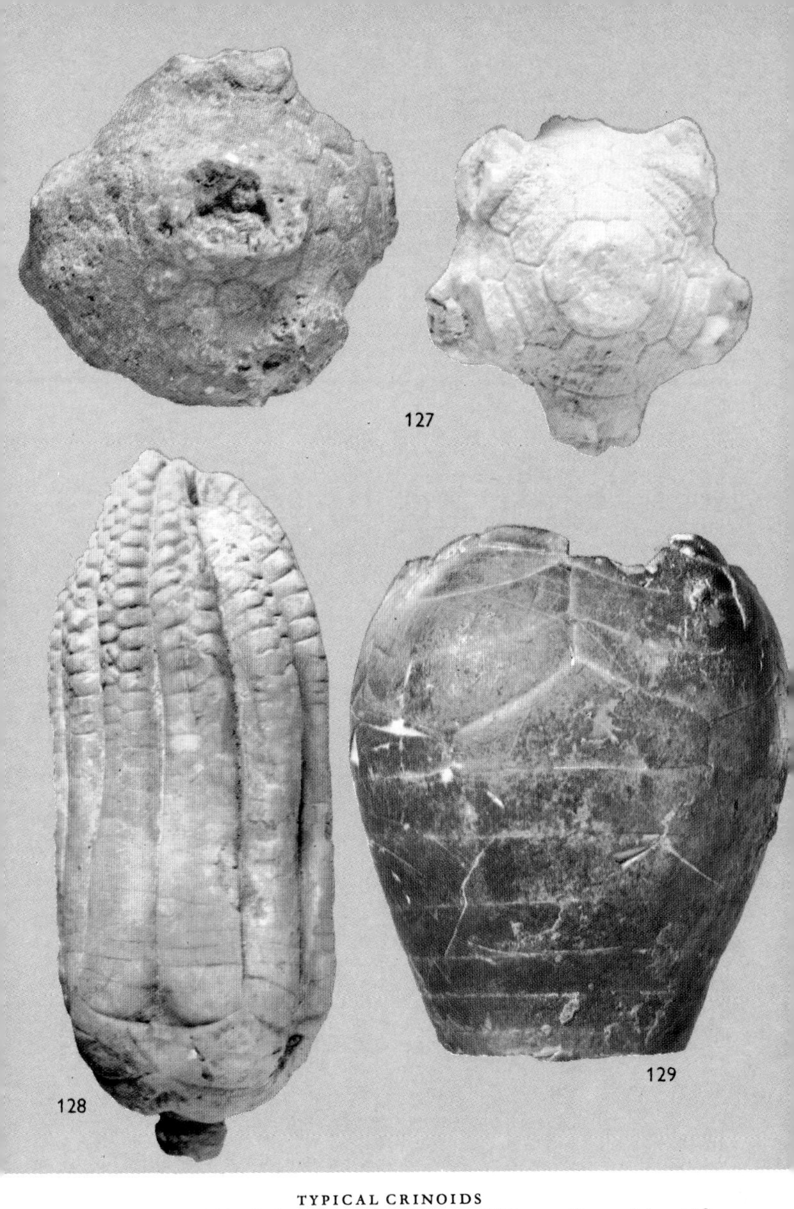

TYPICAL CRINOIDS

127. Camerate Crinoid, Carboniferous Limestone, Clitheroe, Lancashire. ×2
128. *Encrinus,* Muschelkalk (Middle Trias), locality in Germany not known. ×2
129. *Apiocrinus,* Bradford Clay (Middle Jurassic), Bradford-on-Avon, Wiltshire. ×2

THE RECENT REGULAR ECHINOID, *Echinus*

130. Upper surface. $\times \frac{3}{4}$
131. Lower surface. $\times \frac{3}{4}$

CHALK ECHINOIDS

132. *Micraster,* Upper Chalk (Upper Cretaceous), Dartford, Kent. ×1
Left Upper surface
Right Lower surface

133. *Echinocorys,* Upper Chalk (Upper Cretaceous), Highdown, Sussex. ×1
Left Upper surface
Right Lower surface

SANDDOLLARS

134. *Clypeus,* Inferior Oolite (Middle Jurassic), Birdlip, Gloucestershire. $\times \frac{3}{4}$
135. *Clypeaster* (side view), Pliocene, Malta. $\times 1$

MESOZOIC ECHINODERMS

136. *Cidaris,* Corallian (Upper Jurassic), Calne, Wiltshire. $\times 1\frac{1}{2}$

137. *Conulus,* Chalk (Upper Cretaceous), Kingsclere, Hampshire. $\times 1\frac{1}{2}$

138. *Ophioderma* (brittle star), Lias (Lower Jurassic), Seatown, Dorset. $\times \frac{2}{3}$

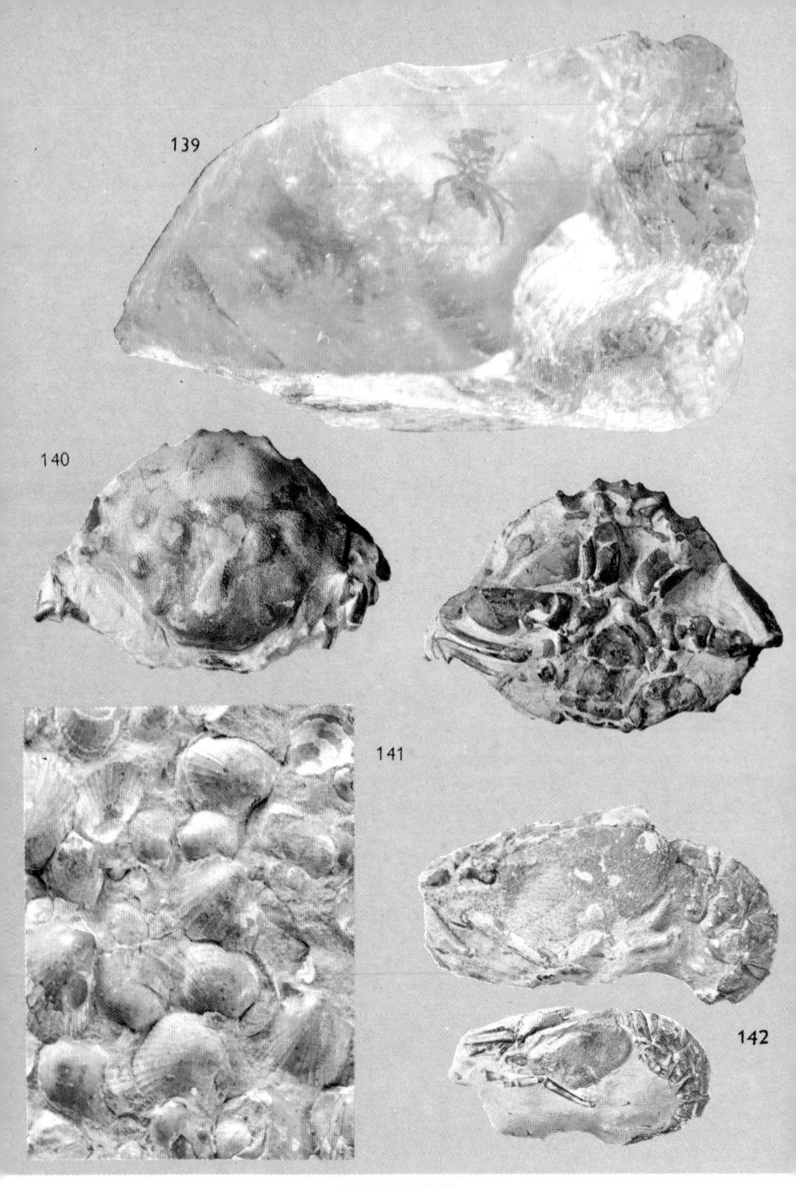

ARTHROPODS

139. Spider in a piece of amber, Oligocene, Baltic Coast. × 2

140. *Xanthopsis*, a crab, London Clay (Lower Eocene), Sheppey, Kent. × 1
Left: Upper surface. Right: Lower surface.

141. *Euestheria* and *Pseudomonotis* (radial ornamentation), Rhaetic Beds, Cotham, Gloucestershire. × 10

142. *Meyeria*, a lobster, Lower Greensand (Lower Cretaceous), Atherfield, I. of Wight.

THE TRILOBITE CALYMENE

143. In normal position. $\times 4$

144. Rolled up. $\times 3$

Wenlock Limestone (Middle Silurian), Much Wenlock, Shropshire.

AGNOSTIDS AND TRINUCLEIDS

145. *Agnostus*, Middle Cambrian, Slemmested, Oslo, Norway. ×1
146. *Trinucleus*, Middle Ordovician, Llandeilo, Carmarthenshire, Wales. ×1

HEADS AND TAILS OF TRILOBITES

147. *Encrinurus,* headshield of an enrolled specimen. $\times 2$

148. *Dalmanites,* headshield showing eyes. $\times 1\frac{1}{2}$
Wenlock Limestone (Middle Silurian), Dudley, Worcestershire.

149. Pygidium of *Scutellum* (*Bronteus*), Middle Devonian, Torquay, Devon. $\times 1\frac{1}{2}$

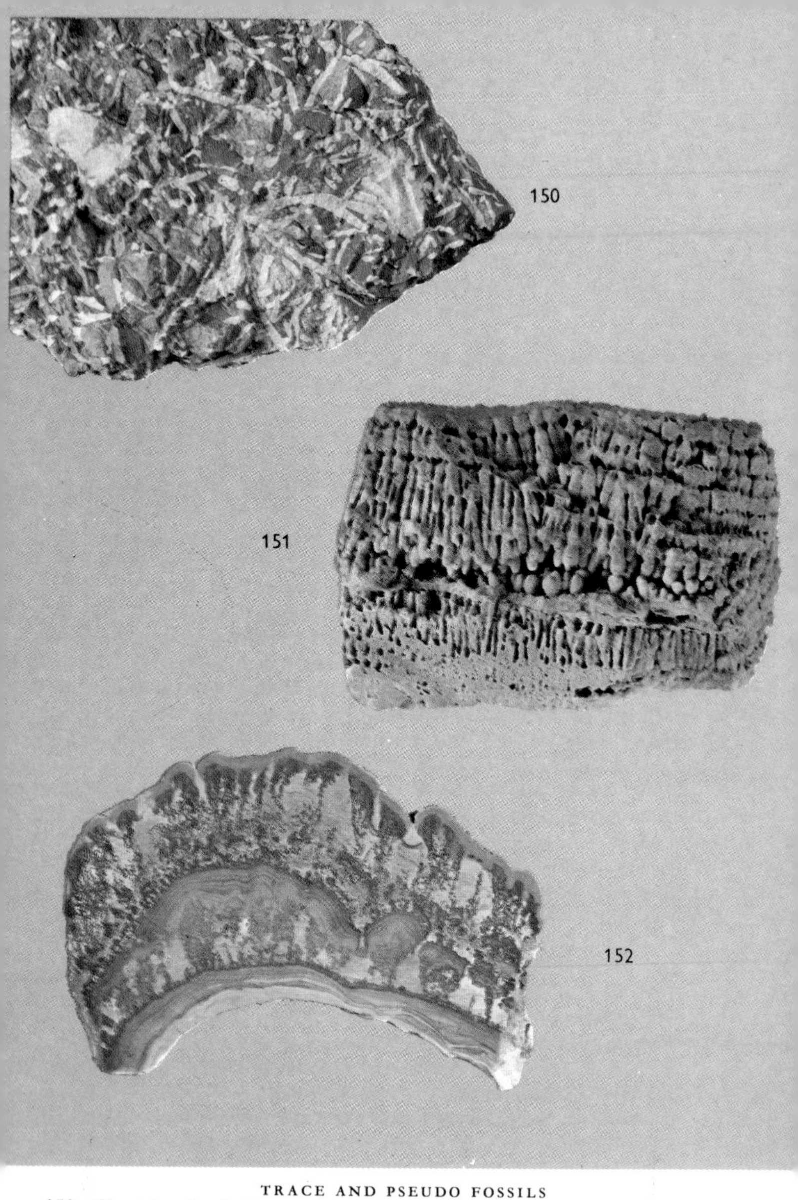

TRACE AND PSEUDO FOSSILS

150. *Chondrites*, Gault (Lower Cretaceous), Henfield, Sussex. × 1
151. Concretionary Magnesian Limestone, Permian, Fulwell, Durham. × 1
152. Landscape Marble, Rhaetic Beds, Bristol, Gloucestershire. × $\frac{1}{4}$

PALAEOZOIC FISH

153. Headshield of *Cephalaspis*, Lower Old Red Sandstone probably Herefordshire. $\times 1$

154. *Osteolepis* in concretion, Middle Old Red Sandstone, Banff, Scotland. $\times \frac{3}{7}$

155. Spine of *Sphenacanthus*, Coal Measures, Tibshelf, Derbyshire. $\times \frac{1}{3}$

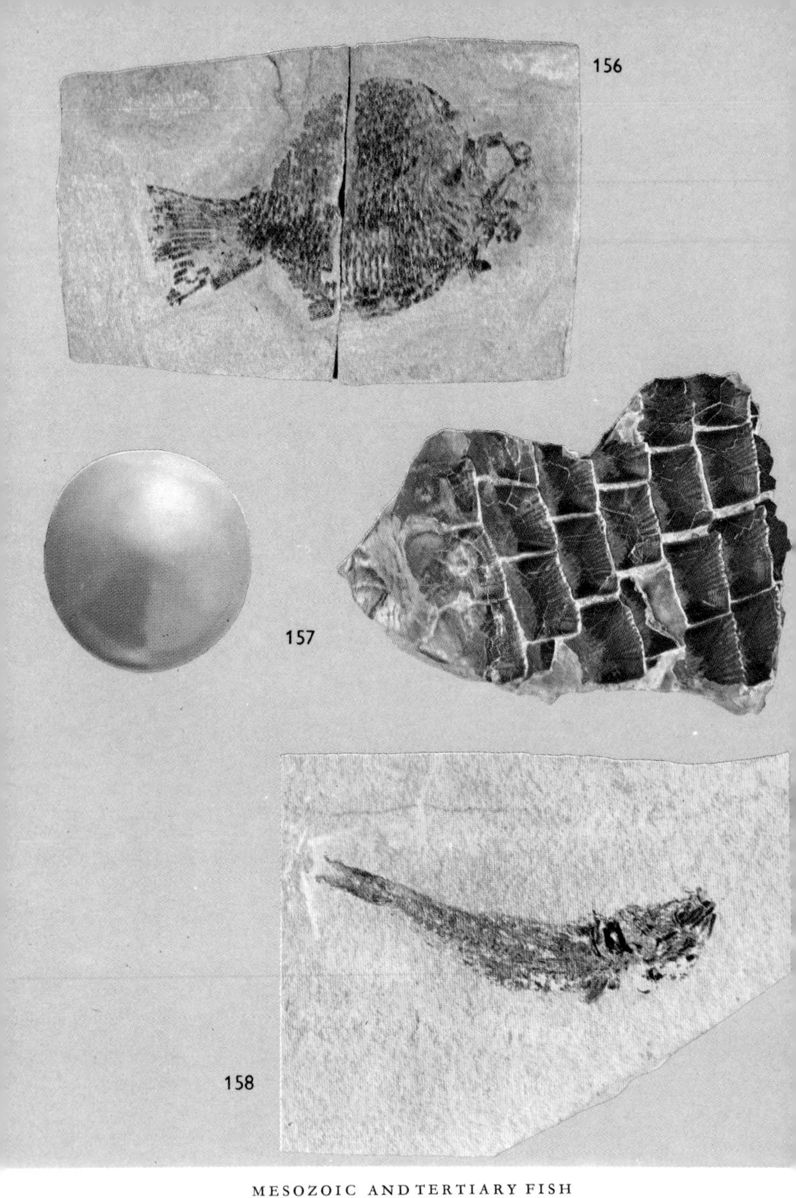

MESOZOIC AND TERTIARY FISH

156. *Dapedius,* Lias (Lower Jurassic), Lyme Regis, Dorset. $\times \frac{1}{2}$

157. Scale and tooth of *Lepidotes*, Weald Clay (Lower Cretaceous), Capel, Surrey. Tooth $\times 2$ Scale $\times \frac{1}{2}$

158. *Clupea*, teleost related to the herring, Eocene, Monte Bolca, N. Italy. $\times 1$

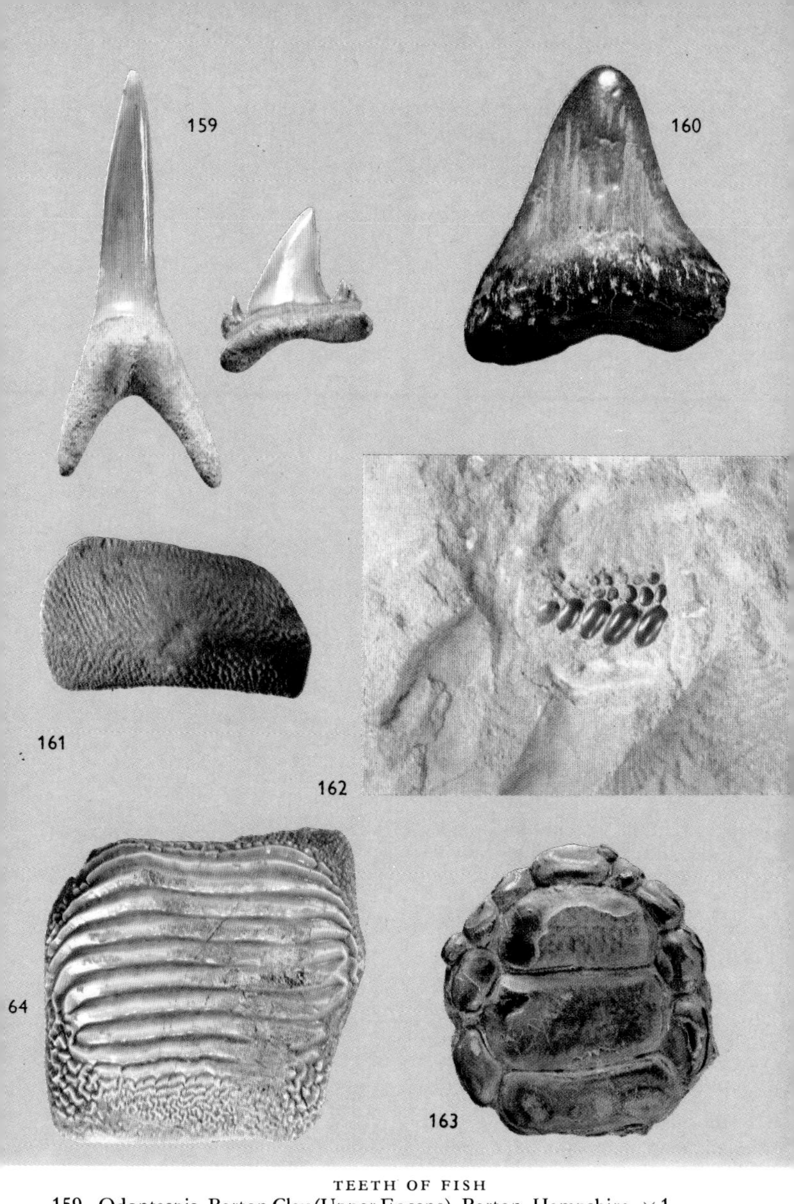

TEETH OF FISH

159. Odontaspis, Barton Clay (Upper Eocene), Barton, Hampshire. $\times 1$
160. *Carcharodon*, Red Crag (Pleistocene), Walton-on-the-Naze, Essex. $\times \frac{1}{2}$
161. *Asteracanthus*, Forest Marble (Middle Jurassic), Atford, Wiltshire. $\times 1\frac{1}{2}$
162. Palatal teeth of *Mesodon*, Middle Jurassic, Stonesfield, Oxfordshire. $\times \frac{1}{2}$
163. *Ptychodus*, a ray, Upper Chalk (Upper Cretaceous), Grays, Essex. $\times {}^3/_2$
164. *Phyllodus*, Tertiary, locality not known. $\times {}^3/_2$

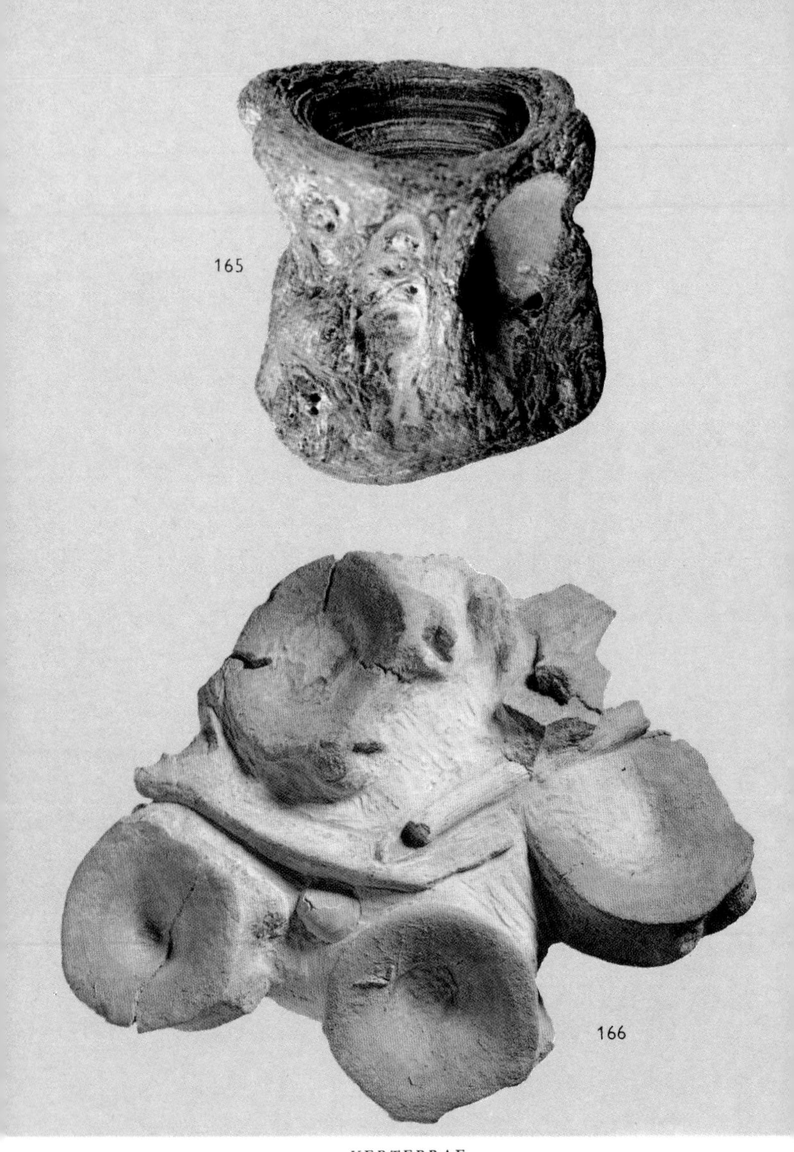

VERTEBRAE

165. Fish vertebra, Bracklesham Beds (Middle Eocene), Bracklesham, Sussex. $\times 2$

166. *Ichthyosaurus* vertebrae, Oxford Clay (Upper Jurassic), Wolvercote, Oxford. $\times \frac{1}{2}$

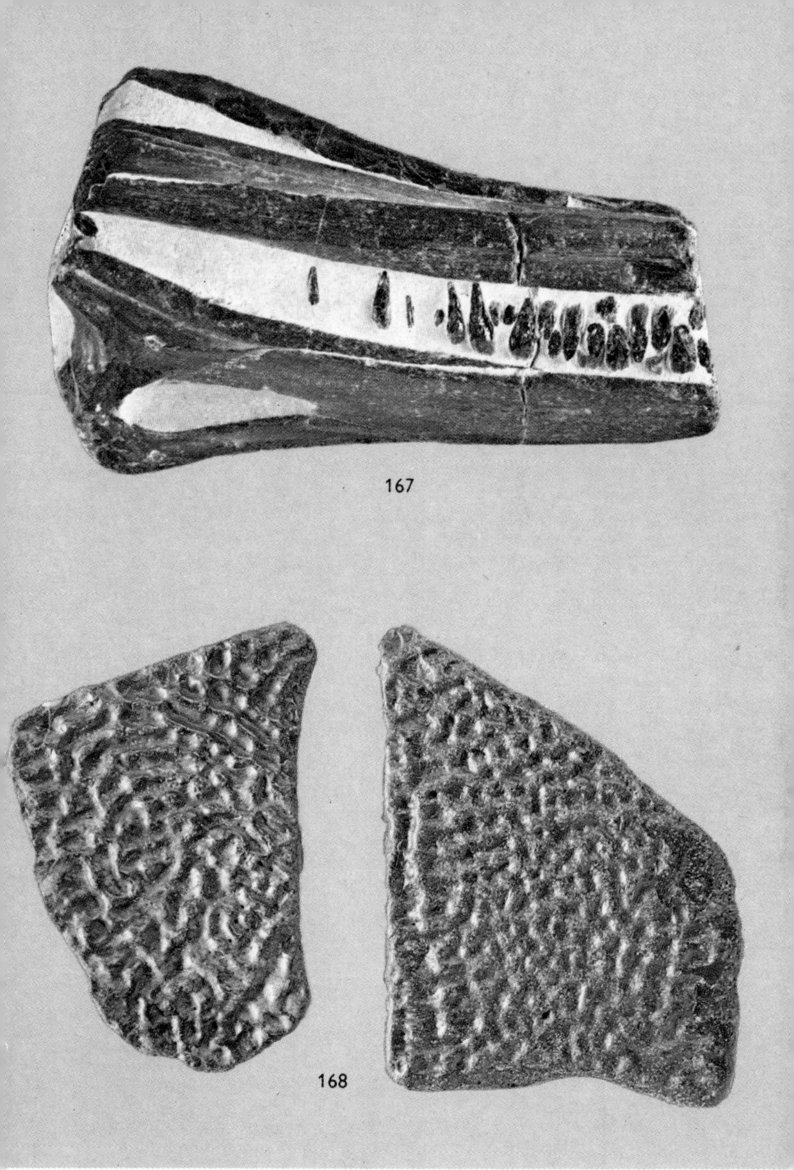

REPTILES

167. *Ichthyosaurus* jaw, Lias (Lower Jurassic), Lyme Regis, Dorset. $\times \frac{1}{2}$

168. Carapace of the turtle *Trionyx*, Hamstead Beds (Oligocene), Isle of Wight. $\times 1$

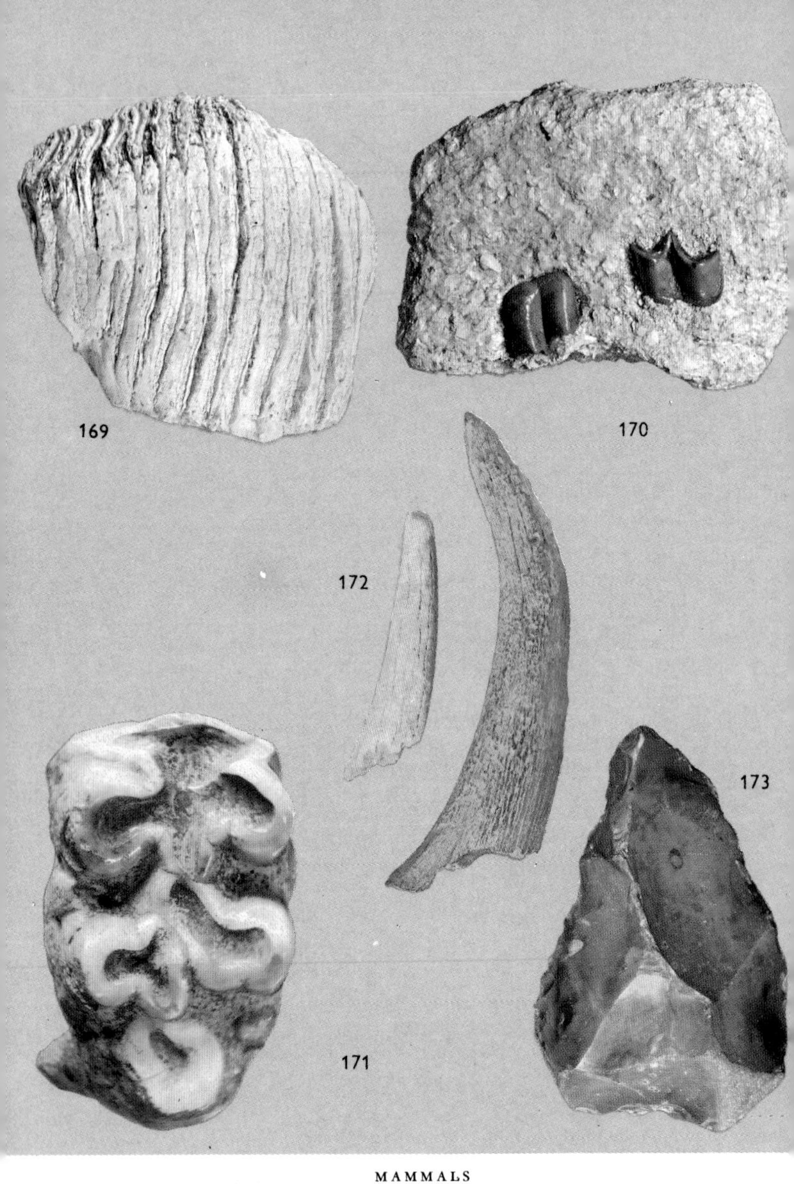

MAMMALS

169. Molar tooth of Mammoth, Pleistocene, Saffron Walden, Essex. ×½
170. *Palaeotherium* tooth, Bembridge Limestone (Oligocene), Bembridge, Isle of Wight. ×2:3
171. *Hippopotamus* tooth, Pleistocene, Malta. ×1
172. Antlers of Deer, Pleistocene, Whitlingham, Norfolk. ×½
173. Acheulian flint implement, Pleistocene, Swanscombe, Kent. ×⅔

PALAEOZOIC PLANTS

174. Algal Limestone, Lower Carboniferous, Randerstone, Fife, Scotland. $\times\frac{3}{4}$
175. *Lepidodendron* stem, Coal Measures, Staffordshire. $\times\frac{1}{2}$
176. *Stigmaria,* Coal Measures, Staffordshire. $\times\frac{1}{2}$

FRONDS

177. *Ginkgo,* Middle Jurassic, Gristhorpe, Yorkshire. $\times \frac{2}{7}$

178. *Pecopteris,* Coal Measures, Staffordshire. $\times \frac{1}{2}$

MESOZOIC AND TERTIARY PLANTS

179. *Equisetites* stem, Hastings Beds (Lower Cretaceous), Balcombe, Sussex. $\times\frac{1}{2}$
180. *Tempskya,* Wealden Beds (Lower Cretaceous), Sandown, Isle of Wight. $\times 1$
181. Seeds of the palm *Nipa,* London Clay (Lower Eocene), Sheppey, Kent. $\times 2\frac{1}{2}$

LEAVES OF ANGIOSPERMS

182. In Tufa, Holocene, locality not known. ×1

183. In pipe clay, Bagshot Beds (Lower Eocene), Alum Bay, Isle of Wight. ×1

the folding of the Variscan geosyncline and the wearing down of the British–Scandinavian land mass.

The *Devonian* picture is relatively simple. In the inter-montane basins and larger depressions on the northern land mass, the *Old Red Sandstone* was deposited. The lower beds in Scotland are pebble beds and coarse sands, but as the mountain chains were worn down, the products of their decay became finer in grade, so the upper beds of the Old Red Sandstone are often fine-grained sands and marls. On this land surface the first land plants developed and in the lower reaches of its rivers and in the shallow waters around the coasts lived the early vertebrates. The Old Red Sandstone of Wales was deposited on delta flats bordering the shallow seas beyond the northern limits of the Variscan geosyncline. Further south in Exmoor, beds of Old Red Sandstone facies interdigitate with the marine Devonian rocks showing that the sea was advancing, retreating, then advancing again across the very flat margins of the northern continent. The same relationship has been proved in borings in the London area and is also exposed along the northern margin of the Ardennes and the Rhenish Schiefergebirge. Further southwards in south Devon, the southern Ardennes and the northern Schiefergebirge, the marine Devonian rocks consist mainly of shales with developments of reef-limestones. The richly fossiliferous Devonian rocks of the Schiefergebirge provide a section across the Variscan geosyncline, for the coarse Lower Devonian quartzites of the Taunus must have been derived from the mid-European island (Map 2).

In *Lower Carboniferous* times the sea transgressed northwards to reach, at times, the Midland Valley of Scotland. In southern Britain shelf limestones, the *Carboniferous Limestone*, were laid down, but in northern England and Scotland the Lower Carboniferous beds are much more variable, for on the margin of the northern landmass, now of very low relief, there were deposited a rapidly varying alternation of marine limestones and shales separated by thick beds of deltaic sandstones, non-marine shales and even coal seams laid down when an increase in the carrying power of the rivers caused deltaic, lagoonal and swamp conditions to spread southwards. In the Ardennes and the extreme north of the Schiefergebirge shelf limestones were deposited. Further southwards the Lower Carboniferous rocks are of a different facies, the *Culm* facies; shales and sandstones laid down in foredeeps, which had developed in front of the growing Variscan folds (Map 3). The Variscan orogenic movements with their intrusion of granites were long drawn out, beginning in late Devonian and early Carboniferous times in the Massif Central, the Vosges, the Black Forest and the Bohemian massif and then spreading slowly northwards, so that the British Isles were not seriously affected until the close of the Carboniferous Period.

The *Upper Carboniferous* rocks of Britain consist of the Millstone Grit, thick alternations of fluviatile sandstones and marine shales, overlain by the Coal Measures. The *Coal Measures* show a repeated *rhythmic* sequence of:

coal seam
seat earth or fossil soil
sandstone
non-marine shales
marine band
coal seam

The land surface must have been of very low relief, for the coal seams originated as peats formed in extensive swamps. An advance of the sea submerged the swamps and a marine band was deposited, but the marine incursions were shortlived. River-borne detritus formed first brackish and then freshwater muds which passed upwards into sandy deltaic flats, which in their turn were colonized by plants. These conditions occurred on both sides of the Wales–Brabant uplands and in the paralic coal basins of Belgium and the Ruhr (Map 4). Certain marine bands, recognizable by their distinctive fossil-content, can be traced from the Scottish, Welsh and England coalfields through north France and Belgium into the Ruhr and indeed into southern Poland.

Finally, in very late Carboniferous times the *Variscan orogenic movements* affected southern England. They are often given the name *Armorican movements*. The Devonian and Lower Carboniferous rocks of Devon and Cornwall were so strongly folded that many of their fossils were obliterated, but fortunately in the Ardennes and the Schiefergebirge the folding was less intense. Further northwards in Britain the folding was but slight, though there was much faulting. Then followed a period of erosion, during which the Coal Measures were worn off the upstanding areas, so that our existing Coal basins date from this period of earth movement some 300 million years ago. In France and in Germany, south of the Rhenish massif, in late Carboniferous times, coals were formed in a number of isolated basins within the Variscan mountain belt, but as the climate gradually became more arid, the luxuriant forests died out.

The geography of Europe was greatly altered as a result of the Variscan movements. The wide fold belt became a land area and a new geosyncline, the *Tethys*, developed along its southern margin (Map 5). In the British Isles, the *New Red Sandstone*, of Permian and Triassic age, was deposited during this period as a result of the erosion of the uplands formed by the Variscan movements. At intervals, parts of the European land mass were submerged by shallow epicontinental seas spreading northwards from the Tethys. The first advance, in Upper *Permian* times, covered much of Germany, so that the continental Rothliegende is there overlain by the *Zechstein*, a marine limestone yielding a fauna of Palaeozoic type. For a short period the Zechstein sea spread into northern England and north-east Ireland to deposit the Magnesian Limestone, but these conditions were of short duration and both in Germany and in Britain the marine limestones pass upwards into thick beds of rock salt and gypsum, deposited from the evaporation of waters too saline for forms of life to survive.

In earliest *Triassic* times, more normal deposition under arid conditions was resumed to produce the *Bunter* Sandstones of Germany and Britain. Again a shallow sea spread northwards from the Tethys into Germany. The fossiliferous *Muschelkalk* was deposited as the middle member of the tripartite Trias of Germany. But the Muschelkalk transgression did not reach the British Isles (Map 6). Again communication with the Tethys was interrupted, so the Muschelkalk is overlain by continental deposits, the *Keuper* of Germany. The European land surface had by now been so greatly reduced in relief, that very probable that the *Upper Cretaceous Transgression* was a world-wide phenomenon, the most extensive submergence of the continental masses known during the geological past. As the transgression advanced, diachronous greensands with a shallow-water fauna were deposited around the shrinking land areas, whilst further off-shore the greensands passed laterally into mud such as the Gault mentioned on p. 15. At the maximum of the transgression, thick beds of chalk were laid down over most of western Europe (Map 10). The Chalk is a unique type of limestone of great purity, extremely fine-grained and of wider geographical extent than any other limestone in the geological column.

Throughout the Jurassic and Cretaceous Periods the Tethyan geosyncline had a complicated history, for it was divided into a number of deep depressions (fosses) by unstable geanticlinal belts which must have formed, at times, lines of island arcs. Fine-grained clays and thin limestones were deposited in the fosses, passing into thicker and much coarser beds towards the geanticlines. On the northern margin of the Tethys great thickness of bioclastic and biohermal limestones were laid down at times, limestones yielding a distinctive warm water fauna of thick-shelled rudistids (p. 149) foraminifera (p. 118), sea urchins (p. 160), etc., a benthonic fauna differing markedly from that living in the epicontinental seas. The Tethyan nektonic fauna included a great variety of ammonites (p. 143), which were able to migrate northwards into Germany, France and Britain and so provide us with the zone fossils needed to correlate and trace time planes through the rapid lithological changes of the Jurassic and Cretaceous beds.

At the close of the Mesozoic Era there was another major change in the geography of Europe. This was not the result of a great orogeny as at the beginning and end of the Upper Palaeozoic Era. No mountain chains were formed in Europe, as at this time in the western United States. A broad uplift of the European land mass caused a regression of the epicontinental seas. In Denmark there are distinctive marine deposits (the Danian Chalk) spanning the time interval between the Cretaceous and the Lower Tertiary Periods. Nearly everywhere else the oldest Tertiary Beds rest with slight unconformity on an eroded surface cut across various horizons of the Chalk. The general pattern was simple (Map 11). In southern Europe a foredeep had formed in front of the growing Alpine chains. In this was deposited the Flysch, sands and shales, partly marine, partly non-marine. In the lakes and other depressions on the land mass to the north accumulated beds, locally containing the remains of contemporary mammals. The *Lower Tertiary* deposits of eastern England, north France, the Low Countries, north Germany and west Denmark, are a varied succession of sands and clays, dominantly marine towards the North Sea Basin, mainly estuarine or freshwater towards the surrounding land masses. The succession of the Lower Tertiary rocks is everywhere complicated in detail, for there were a number of minor advances and retreats of the margins of the seas. Intermittently communication opened along the line of the English Channel, round the Brittany land mass, through the Gulf of Aquitaine to the Tethys. When this occurred members of the distinctive Tethyan fauna were able to spread into the North Sea Basin.

In north-west Britain conditions were very different. Volcanic activity produced the great lava plateaus of Antrim and the western Isles of Scotland. These are but the southern part of the *North Atlantic Volcanic Province*, which included the Faeroes, Iceland and eastern Greenland. Volcanic activity was not continuous, for at intervals in the lava pile there are deeply weathered fossil soils in which are preserved trunks of trees and other plant remains. Activity in the British Isles probably ceased by the end of the Lower Tertiary, but has not yet done so in Iceland. In Upper Tertiary times there was much volcanic activity on the continent of Europe, notably in the Massif Central of France, and in parts of Germany. Here again fossiliferous sediments are locally to be found interbedded with the sheets of lava and volcanic ash.

At the close of the Lower Tertiary occurred the main folding of the great thickness of rocks which had accumulated since the beginning of the Mesozoic in the Tethyan geosyncline. The intensely folded rocks drove northwards to form the Young Mountain Chains of Europe, the Pyrenees, the Alps and the Carpathians. In France, Germany and southern Britain the ripples of the *Alpine Orogeny* caused some gentle folding, such as the depression of the Paris Basin and the broad rise of the Weald separating the London and Hampshire Basins. Throughout the Miocene Period the British Isles seem to have been a land area, for marine beds of this age are only found on the east side of the North Sea. Parts of north-western France were, however, submerged by a shallow sea in which were deposited richly fossiliferous shelly sands, the 'faluns'. In Pliocene times the sea returned to East Anglia to deposit beds of similar lithology, the well known 'Crags'. A depression extending, in front of the Alpine Chains, was invaded by the sea during the early part of the Miocene Period, but regression soon followed, and the depression became a chain of brackish water basins, which gradually became silted up and reduced to the present Caspian Sea and the Sea of Aral (Map 12).

Throughout the Mesozoic and most of the Tertiary era climatic conditions in Europe were warm, often subtropical, but now came a rapid refrigeration, producing glaciers on the mountains of Scandinavia, the British Isles, the Alps and the Pyrenees. As the mountain glaciers grew they coalesced and spread out across the lowlands to form great ice sheets (Map 13), with wide belts of barren tundra and arid steppe to the south. The story of the Pleistocene Period is extremely complex, for it comprises a number of glacial periods separated by interglacial periods, when climatic conditions may well have been like the present. Each advance of the ice ploughed up any unconsolidated material, so that our record of the older interglacial periods is very incomplete, for their deposits have escaped destruction at only a very limited number of localities. Finally the ice sheets shrank back into the mountains and by the beginning of the Holocene the present geography of Europe had come into being.

IV. THE CHIEF GROUPS OF FOSSIL ORGANISMS

1 PROTOZOA

The protozoans are the simplest members of the Animal Kingdom, for they are built up of only one cell. This consists of a semi-liquid living substance called *protoplasm*, in which are one or more denser nuclei. Food particles, such as diatoms, can be absorbed either by any part of the body or through thread-like mobile extensions of the body known as *pseudopodia*.

The majority of the protozoans are microscopic in size and therefore their study has been dependent on the development of suitable means of magnification. Again the majority of protozoans are without hard parts and so have not been recognized with certainty in the fossil state. Certain groups of protozoans do, however, secrete skeletons, often of great beauty and complexity. They are to be found in a great variety of sedimentary rocks, sometimes in such abundance as to be important rock formers, at many other horizons they are important zonal indices and, to a lesser extent, indicators of the conditions under which the rocks containing them may have been deposited. The micropalaeontologists of the oil industry are mainly concerned with the study of the vertical and horizontal distribution of fossil protozoans together with other microfossils including the early growth stages and, in some cases, fragments of the adult forms (see p. 24) of more advanced organisms.

The Protozoans are subdivided into the following classes:

(*a*) Flagellata (or Mastigophora).
(*b*) Sarcodina (or Rhizopoda).
(*c*) Sporozoa.
(*d*) Ciliata.

The last two classes are without fossil representatives.

1*a* The **Flagellata** have a definite form, for the protoplasm is surrounded by a firm membrane. They take their name from their *flagellae*, fine whip-like threads of protoplasm which are used for locomotion. In recent years also, with the development of suitable methods for their extraction and using magnifications of scores or hundreds of times, it has been shown that many sedimentary rocks contain *silicoflagellates* or *dinoflagellates*, with either a siliceous or a toughened organic skeleton. But the study of organisms a tenth of a millimetre or less, often very much less, in diameter is clearly a matter for the specialist.

1*b* The **Sarcodina** include two important orders of fossils, the Foraminifera and the Radiolaria.

The **Foraminifera** are mainly marine, though some can live in brackish water. The majority of the marine forms live near the sea floor, others are part of the plankton of the surface waters. They secrete a skeleton which may be chitinous, calcareous or, in the arenaceous Foraminifera, composed of small sand grains, mica plates, sponge spicules, shells of other Foraminifera, etc., bound together with a calcareous or ferruginous cement. Their remains are accumulating today in sufficient abundance in the ocean depths to form extensive deposits of ooze, made up largely of the planktonic form *Globigerina*. Benthonic forms are more numerous in shallower waters, but their remains are swamped by the supply of land derived detritus. The larger Foraminifera, a quarter of an inch or more in diameter, occur in tropical waters, especially around coral reefs; and indeed Foraminifera play a not insignificant part in the make up of a 'coral reef'. Many other lime-secreting organisms reach their maximum size in hot shallow seas, for unlike most other mineral salts, the solubility of calcium carbonate decreases with rise of temperature and therefore in such a setting, the sea waters can easily become oversaturated, so that the excess lime is available either to be chemically precipitated or to be secreted by organisms.

The shell or *test* of Foraminifera is made up of a number of intercommunicating chambers arranged in a wide variety of ways, ranging from a straight line (*Nodosaria*), a double line (*Textularia*), a plane spiral (*Operculina*), a helicoid spiral (*Globigerina*) (Fig. 9) to very complex arrangements as in *Alveolina* or *Nummulites* (No. 17). The outside of the test normally shows numerous perforations for the passage of the *pseudopodia*, especially in the *vitreous forms*, which have a glassy appearance, but in the *porcellaneous* foraminiferans, white and opaque in reflected light, the shell is imperforate and the pseudopodia pass out through one or more apertures. The outer walls of the larger Foraminifera such as *Nummulites* are thickened by the development of a supplemental skeleton, carrying a distinctive ornamentation.

The larger Foraminifera, that is those visible to the naked eye, can be collected from unconsolidated sediments, either by hand or by concentrating them in a sieve. But when the sediments are more compacted it is necessary to break down cohesion and convert the shale or mudstone into a mud. This can be done by heating it in a solution of hydrogen peroxide or by dropping heated fragments into cold water. The mud is carefully decanted away, to leave a clean residue of sand grains and small fossils. This residue is then examined under a hand lens or low-power binocular microscope and the Foraminifera picked out with forceps, or in the case of the smaller ones, on the tip of a moistened sable brush. The shells of Foraminifera are very thin, so that they are easily abraded. Any assemblage of Foraminifera obtained by the methods given above will include individuals showing every variety of wear from undamaged specimens with the ornamentation of the outer wall complete to those in which the outer wall has been worn right through to reveal the arrangement of the chambers. It is partly owing to their minute size that Foraminifera are of such value to the micropalaeontologist, for uncrushed specimens can be obtained from the small rock chips brought up from an oil well and studied under a binocular microscope. Thin sections of sediments made in the normal way for examination under the petrological

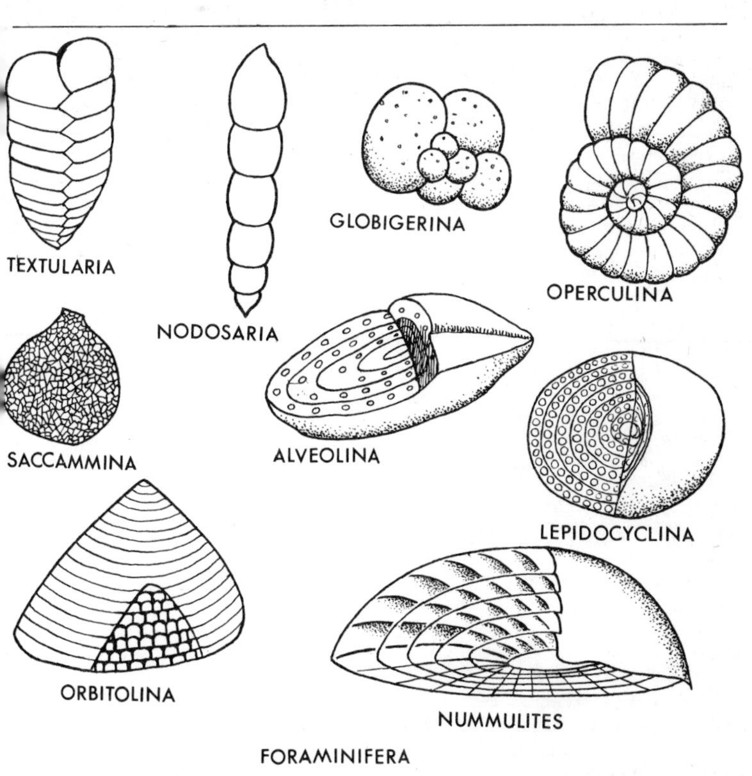

Fig. 9. *Representative protozoans, drawn to different magnifications.*

geologists often use the term Nummulitique for the Lower Tertiary, for this was the period when Foraminifera reached their maximum importance in the rocks of Europe. The palaeogeography of Europe during Upper Tertiary times was not very favourable to the requirements of Foraminifera, especially the larger foraminiferans, but in other parts of the world they occur in abundance, and are of great value to the micropalaeontologists of the oil industry.

The **Radiolaria** secrete a skeleton of silica, an intricate skeleton often of the most delicate tracery (Fig. 15). They are minute pelagic forms, flourishing in particular in the surface waters of the warmer parts of the Pacific and Indian Oceans. After death their minute skeletons sink down to form extensive deposits of *radiolarian ooze* on the deeper parts of the ocean floors. Radiolarian oozes are found at depths of between 2,000 and 4,500 fathoms. The calcareous shells of Foraminifera and coccoliths (p. 189) pass into solution at depths greater than 2,000 fathoms and whilst the siliceous shells of the radiolarians can withstand a much greater pressure of water, even they are eventually dissolved. Below about 4,500 fathoms, the deepest parts of the oceans are floored with extremely fine-grained Red Clay containing only manganese nodules and teeth of sharks.

At a limited number of horizons in the Lower Ordovician rocks of Southern Scotland, the Lower Carboniferous rocks of south-west England and central France, the Mesozoic rocks of the Alps, there are beds of tough splintery chert, seen to be full of Radiolaria when examined in thin-section under the microscope. It is unlikely from the nature of the surrounding beds that these *radiolarian cherts* were deposited in as deep water as are the modern radiolarian oozes. Rather they were formed in troughs of moderate depth, troughs both sheltered from the supply of land-derived detritus, and whose surface waters were probably particularly suitable for the development of radiolaria. Hence the unusual accumulation of radiolarian-rich deposits. On the other hand, the unconsolidated *radiolarian earths* of Miocene age that occurs on Barbados and other West Indian Islands are most probably uplifted deep-sea deposits.

2 PORIFERA

The Animal Kingdom can be divided into two Sub-Kingdoms – the Protozoa (*one-celled*) and the Metazoa (*many-celled*). The Porifera (*pore-bearing*), the Sponges, are the simplest of the metazoans. The sponges are aquatic animals, mainly marine. Their architecture and mode of life are simple. Their body walls are supported by an internal skeleton and are porous, being traversed by many passage-ways lined by collar cells, each bearing a single whip-like *flagellum*. The constant movement of the flagella causes currents of water to flow through the passage-ways into the central hollow or *cloaca*. As the water moves through the walls of the sponge, oxygen and nutritive particles of very minute size are absorbed. They have no separate mouth, stomach or anus. Their skeleton is composed either of spicules of calcium carbonate or of opaline silica or spongin, a tough flexible organic substance. There is considerable variation (Fig. 10) in the form of the spicules. It is only those sponges with interlocking spicules forming a rigid skeleton (No. 19) that are likely to be preserved as fossils. When isolated spicules are embedded in spongin, the spongin will disappear after death and a loose aggregate of spicules is all that will be left.

Sponges vary considerably in form, ranging from colonial mat-like encrusting types to rigid solitary individuals 2–3 feet in height and width. *Entobia* (No. 22) is unusual, for it burrowed into the calcareous shells of lamellibranchs. The ideal conditions for such benthonic forms were in clear current-swept water.

The phylum of the Porifera is divided into three classes. The Calcispongea with calcareous spicules, further subdivided into the thin-walled Sycones, e.g. *Barroisia* (No. 20) and

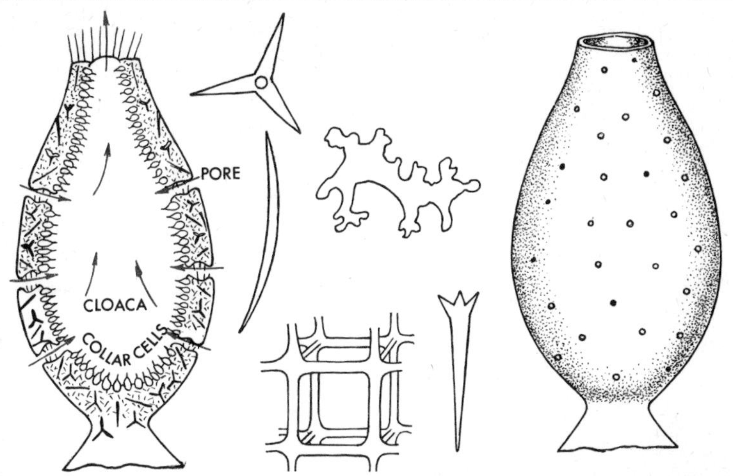

Fig. 10. *Sponges. On left – Cross-section of a sponge, the arrows show direction of water currents. In centre – Different types of sponge spicules. On right – Outer surface of sponge shown on left.*

the thick-walled Pharetrones, e.g. *Raphidonema* (No. 21), the Hyalospongea with siliceous spicules, e.g. *Cephalites* (No. 6) and the Demospongea. The Demospongea include the 'commercial' sponges with thin walls made only of spongin and the lithistids, with irregularly shaped siliceous spicules producing a strong skeleton, e.g. *Doryderma* (No. 23).

Sponge spicules have been recorded from the Pre-Cambrian rocks of northern France and the Cambrian rocks of South Wales. At a limited number of horizons sponges occur in abundance – horizons such as the Lower Cretaceous Sponge Gravels of Farringdon in Berkshire, rich in calcareous sponges, and the lithistid-bearing Upper Greensand of the Haldon and the Blackdown Hills in Dorset and Devonshire. In the Upper Jurassic limestones of the Jura Mountains of south-western Germany sponges play an important part in building sponge-coral reefs. Sponge remains are not uncommon in certain limestones, such as parts of the Carboniferous Limestone and the Chalk. The opaline silica of their spicules is soluble in alkaline water and may be reprecipitated to help form nodules of chert and flint. For example, the *Cephalites* shown in No. 6 is a ferruginous pseudomorph of an originally siliceous sponge.

The **Archaeocyathids** or pleosponges are an extinct group occurring only in rocks of Lower and Middle Cambrian age. In these they have a world-wide but sporadic distribution. In certain areas of Australia, Antarctica and the United States they form thick and extensive reef-limestones. Their systematic position is uncertain. Conical in form, but quite small with a

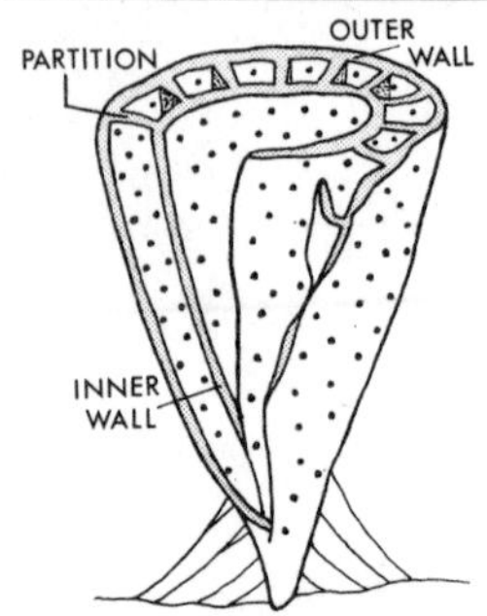

Fig. 11. *An Archaeocyathid. Pores shown in black.*

maximum height of 4 inches, they have an inner and an outer wall connected by radially arranged partitions (Fig. 11). To this extent, they resemble the corals, but both the walls and the vertical partitions are strongly porous – a sponge-like characteristic. They have been grouped with the sponges, the corals, the bryozoans and the algae. Others regard them as sufficiently distinct to form a phylum on their own, and if so the name archaeocyathids (*ancient cups*) is preferable to pleosponges (*true sponges*).

3 COELENTERATA

The Coelenterata (corals, jelly fish, sea anemones, etc.) are characterized by radial symmetry and a sac-like digestive cavity, hence their name derived from the Greek *coel*, hollow and *enteron*, gut. An alternative name, not in common use, is Cnidaria (*knide*, nettle) after their possession of stinging cells used to immobilize prey, which is then caught by the tentacles, carried to the central mouth and so into the digestive cavity. Contact with the larger jelly fish can be painful, indeed dangerous, even to human beings. Coelenterates are aquatic organisms, mainly marine and the majority are colonial. Many are benthonic, the others are pelagic, whilst in some coelenterates there is an alternation of attached and medusoid (jelly fish-like) generations. Only certain coelenterates secrete hard skeletons.

The chief subdivisions of the Phylum are:

CLASS	SUBCLASS
Hydrozoa	
Scyphozoa	
Anthozoa	Alcyonaria
	Tabulata
	Zoantharia

3*a* Hydrozoa

The majority of the **Hydrozoans** are either soft bodied or have, at best, a chitinous skeleton. Hence they are very rarely preserved fossil. Two subclasses, however, the Milleporida and the Stylasterida, secrete calcareous skeletons and play an important part in the building of modern coral reefs. They occur much less frequently in reef-limestones as far back as Cretaceous times. Their minutely porous skeletons are penetrated by straight-sided tubes at right angles to the surface of the colony. The tubes are of two sizes, the larger one, *gastropores*, housing short feeding polyps, the smaller ones, *dactylopores*, longer slender mouthless polyps armed with stinging cells, both for protection and for food-catching.

The **Stromatoporoids** are an extinct group, whose systematic position and status is debatable, but many authorities group them with the hydrozoans. Marine colonial forms with a calcareous skeleton, they range from the Cambrian to the Cretaceous, but are most important as reef-builders in limestones of Silurian and Devonian age in

England, Belgium and the Isle of Gotland in the Baltic. Certain forms occur as rounded masses, but more commonly the stromatoporoids form thin laminated sheets with irregularly arranged vertical pillars at right angles to the laminae. Their detailed structure is well seen in polished specimens (No. 24). Weathered specimens often show on their upper surface a characteristic pattern of low rounded elevations from which slight grooves radiate. The reef-building stromatoporoids of the Palaeozoic limestones are found associated with algae, rugose and tabulate corals, though they seem to have been able to survive in more turbulent water than the corals. In the Mesozoic rocks stromatoporoids are much less common and occur not in reef-limestones, but in oolitic limestones.

3*b* Scyphozoa

The **Scyphozoans**, the jelly fish, are entirely soft-bodied and are therefore found as fossils only under the most exceptional conditions of preservation. They do, however, range back as far as the Cambrian and have been claimed from rocks of late Pre-Cambrian age. Impressions of jelly fish as large as 16 inches in diameter have been found in the Lithographic Stone (p. 115). The *Conularids* are a problematical group of fossils, occasionally found in rocks of Palaeozoic age. Their steeply pyramidal shell, quadrilateral in cross-section is ornamented by a delicate network of lines (Fig. 12). The shell is thin and composed of chitinophosphatic material. Young forms show signs of attachment by the apex of the pyramid, but in adult life the attachment was severed and the conularids are believed to have behaved as medusoids, floating with the apex of the pyramid upwards and tentacles hanging downwards.

3*c* Anthozoa

The **Anthozoans**, are by far the most

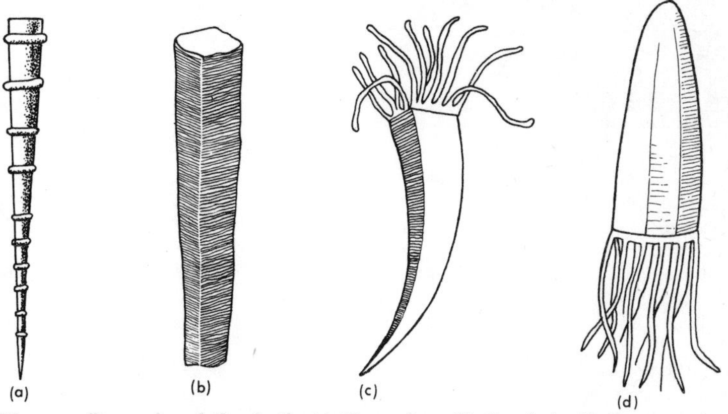

Fig 12. *Pteropods and Conularids. (a) Tentaculites, (b) Conularia, (c) Reconstruction of a juvenile conularid (after Kiderlen), (d) Reconstruction of an adult medusoid conularid (after Kiderlen).*

important class of the coelenterates to the geologist. Their characteristic features are (i) there is no medusoid stage and (ii) the digestive cavity beneath the slit-like mouth, is divided by a number of vertical partitions or *mesentaries*.

The **Alcyonarians**, or **Octocorals**, with eight mesentaries and eight tentacles, range back to the Trias, but are not an important fossil group. Many of them, including the sea pens and some of the horny corals, had a skeleton that was only partially calcified or was composed of loosely interlocked calcareous spicules. Certain octocorals such as *Heliopora* and *Tubipora* (the organ-pipe coral) are important in the make up of modern coral reefs.

The **Tabulata** are an extinct group that are entirely restricted to the Palaeozoic rocks. They are colonial forms with a calcareous skeleton composed of numerous tubes or *corallites*. The corallites are crossed by closely spaced horizontal partitions or *tabulae*. The inner walls of the corallites are usually smooth, but in some genera very short vertical *septa* or septal spines are developed. Whilst the individual corallites are only a fraction of an inch in diameter, hundreds or thousands of them may be present in a fully developed corallum, which may grow upwards from a narrow base to a yard or more in diameter at its upper surface. The arrangement of the corallites varies. In *Favosites* (the honeycomb coral) (No. 25) the corallites are parallel and closely packed, so that they are polygonal in cross-section, and the thecal walls are traversed by mural pores. In *Halysites* (the chain coral), the corallites are elliptical or circular in cross-section (No. 26). In *Syringopora* the trumpet-shaped corallites are in contact only at their lower ends. In *Heliolites* the corallites with well developed septal spines are set in a fine-grained network (No. 27). Cross-sections of the 'feather coral', *Pachypora* (No. 28), cut the thick walled curving corallites at differing angles. Certain genera such as *Vaughania* or *Michelinia* are much smaller and disc-like with few open corallites. The aptly named *Pleurodictyum problematicum*, a marker fossil in the Lower Devonian rocks of the Ardennes and the Rhineland, seems to have grown round a worm. It is usually found as moulds in sandstone, so the short bars are really mural pores infilled with matrix, whilst the thecal walls have been dissolved away.

The oldest tabulate corals are found in the Lower Ordovician rocks of the United States. In Europe and Asia they first appear in beds of slightly later date. In the reef-limestones of Silurian age in many parts of Europe, the tabulate and the rugose corals are of equal importance as reef-builders, but in the Devonian and Carboniferous limestones the rugose corals progressively outnumber the tabulates. Certain Mesozoic and even Eocene corals have been referred to the Tabulata, but the evidence is not too strong and it is now accepted that the tabulates did not survive the close of the Palaeozoic era.

As with other extinct groups, such as the archaeocyathids and the stromatoporoids the systematic position of the Tabulata is uncertain, as is also their relationship to the other main groups of corals.

The subclass of the **Zoantharia** includes the sea anemones, without hard parts, together with many corals. One distinction from the alcyonarians is the absence of spicules in the skeleton of

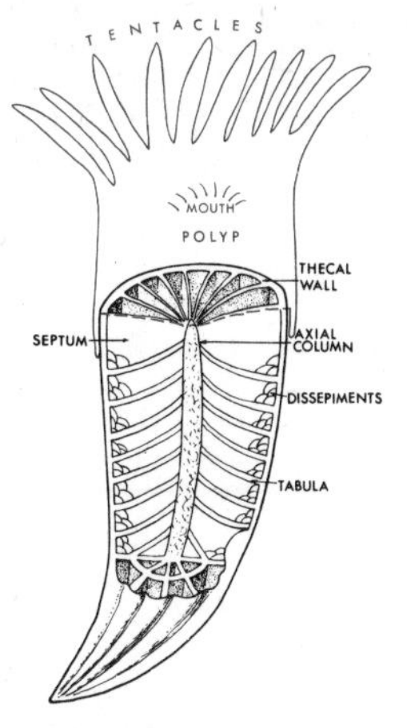

Fig. 13. *Morphology of a solitary coral. Soft parts in red – Hard parts in black. N.B. This is a generalized form.*

the corals. Also beneath the mouth of the zoantharians there is an elongate gullet leading the digestive cavity.

The hard parts of a zoantharian consist fundamentally of a basal disc, from which rise an outer wall or *epitheca* whilst vertical septa were secreted between the *mesentaries*. Gently arched *tabulae* extend across the theca or cup and these may be strengthened near the thecal walls by the development of many small strongly curved *dissepiments*. An *axial column*, sometimes of complex structure, may be formed in the centre of the thecal cup (Fig. 13).

The zoantharian corals are divided into two main groups. The Rugose or Tetracorals, entirely Palaeozoic, and the Scleractinia or hexacorals ranging from Mesozoic to Recent. The septa of certain of the Rugosa are inserted in quadrants, hence the name **Tetracorals**. The earliest rugose corals, found in the Ordovician rocks of the United States, show considerable resemblance to the early tabulates. Rugose corals first became abundant in the Silurian rocks and by Upper Palaeozoic times they were the dominant group of corals, but they did not survive into the Mesozoic.

The **Rugose** (*wrinkled*) corals exhibit great morphological variation. As well as being either simple or compound with the corallites packed in different ways, all their major structural elements – septa, tabulae, dissepiments, axial column – can be modified in various ways. Common genera found in the Silurian rocks are *Acervularia*, a compound form with polygonal corallites, whose septa are thickened half way along their length to produce an 'inner wall' within the thecal wall (No. 30). *Omphyma* is a large solitary coral with the septa prominent only near the thecal walls. Compound corals of Devonian age are *Acervularia*, *Phillipsastraea* without well-developed thecal walls so that the septa are *confluent* running from one corallite to the next (No. 31), whilst the common Devonian solitary corals include the slipper coral *Calceola* (No. 41), *Heliophyllum* with short cross bars on the septa and *Cyathophyllum* with numerous septa extending almost to the centre of the corallite. The cyathophyllids range into the Carboniferous rocks where they may be several inches in diameter (No. 32). *Caninia* is an even larger solitary coral sometimes up to a foot in length and often curved, when the corallite had fallen over to one side and then continued growing upwards. *Zaphrentis* with the ring of widely spaced stout septa interrupted by gaps or *fossulae* (No. 33) is another typical solitary coral of the Carboniferous Limestone. The upper beds of the

Carboniferous Limestone yield a distinctive coral fauna including *Dibunophyllum* (No. 35) and *Aulophyllum* (No. 34), solitary corals with well developed and characteristically shaped axial columns, and the compound corals *Lithostrotion* with a stout flattened axial column and *Lonsdaleia* (No. 40) with a well-developed ring of dissepiments so that the septa do not reach the thecal wall. Corals as well as being important rock formers, are of zonal importance in subdividing the limestones of Carboniferous age.

Precise identification of the rugose corals is dependent on finding specimens showing the detailed arrangement of septa, axial column, dissepiments, etc. These may be etched out by natural weathering. If suitably weathered specimens cannot be found, it may be necessary to grind down the corals to produce transverse sections, but all too often one finds that the skeletal details have been obliterated by recrystallization. Occasionally one finds corals that have been silicified and then if they have not been etched out by normal weathering, the matrix can be removed by treatment with weak acids.

The Scleractinia or **Hexacorals**, with septa inserted in sextants, instead of the quadrants of the Rugosa, have many of the morphological features of the rugose corals. The septa of the hexacorals are, however, often *exsert*, that is they extend down the outside of the thecal walls. The compound hexacorals show the same kinds of packing of the corallites as are found in the tabulate and rugose corals. In addition, some hexacorals are meandroid in habit with the corallites arranged in linear series surrounded by strong walls (No. 36).

The first hexacorals are known from rocks of Middle Triassic age. Their relations to the rugose corals are uncertain. There are two possibilities: (i) that all the corals had a common ancestor in Cambrian times and that whilst the rugose corals soon developed hard parts, the hexacorals did not do so until after the rugose corals had passed into extinction; (ii) that the hexacorals are the descendants of the tetracorals. Future discoveries of corals in rocks of latest Permian or early Triassic age may provide definite evidence to solve this uncertainty.

By late Triassic times the scleractinian corals had a world-wide distribution. They were important reef formers at numerous horizons in the Jurassic and Cretaceous rocks, especially along the Tethyan (p. 115) belt. Well-known Jurassic hexacorals are the solitary coral *Montlivaltia*, and the compound corals *Thecosmilia* (No. 36), *Thamnasteria* with confluent septa and *Isastrea* (No. 37). The solitary form *Parasmilia* is not too uncommon in the normal Chalk, whilst *Dendrophyllia* (No. 38) is an important form in the Danian Chalk. In early Tertiary times, many of the Mesozoic hexacorals died out and were replaced by the families that are dominant today. Corals are but rarely found in the Tertiary rocks of northern Europe. Some coral reefs occur in the early Tertiary rocks of the Tethyan belt, but later in Tertiary times the reef building corals withdrew to their present intertropical distribution. This was probably a response to the lowering of sea temperatures due to the onset in higher latitudes of the Pleistocene glaciation.

3*d* Corals and Coral Reefs

Recent coral reefs are not composed entirely of corals nor are recent corals restricted to reefs. Coralline algae, such as *Lithothamnion* are just as important as corals in building the framework of the reef, whilst in addition bryozoans and Foraminifera are important in binding the reef together. Coral reefs are inhabited by a specialized reef fauna including thick-shelled molluscs, sea urchins and coral-eating fish. All these help in the making of the reef. The reef-building (*hermatypic*) corals and algae can only flourish vigorously in sea water of normal salinity, in depths not greater than 25 fathoms and in waters whose temperature falls below 18°C (65°F) for only short periods. Therefore the modern coral reefs are restricted to the warm-water coasts of the intertropical belt. Other (*ahermatypic*) corals, mainly solitary forms, are widely distributed in colder seas and often at much greater depths. For instance *Caryophyllia* may be found at extreme low tide in pools along the coasts of Devonshire and Brittany. Off the coasts of Norway great sheets of the branching coral *Lophohelia* have been recovered from as much as 300 fathoms, whilst corals have been dredged from the ocean abysses in depths of thousands of fathoms.

It is assumed, though it cannot be definitely proved, that the requirements of the massive colonial rugose and tabulate corals of the Palaeozoic Era were similar to those of modern reef builders. Fossil coral reef-limestones have, however, a much greater latitudinal range than modern coral reefs. Reefs of both Silurian and of Carboniferous age have been recorded from as far north as northern Siberia and as far south as Australia, and Cretaceous coral reefs from southern South America to the Mediterranean belt.

If our premise above is correct, the temperature belts at these periods must have been much less sharply defined than today and indeed there may have been no really cold areas. In detail the Silurian reefs of England, the Isle of Gotland and the central United States, the Devonian reefs of the Ardennes and Canada, the Carboniferous reefs of many parts of Europe, the Upper Jurassic reefs of England, France and south-west Germany, the late Cretaceous reefs of Denmark show many of the features of modern coral reefs. The central part of the reef is composed of compound corals in the position of growth, together with stromatoporoids, sponges and bryozoans, whilst on the margins of the reef mass, the corals and other organisms have been displaced by wave action and are lying at any angle. The reef-limestones also yield a specialized fauna of brachiopods, crinoids, etc.

But fossil corals are not restricted to the reef facies. They occur, though not so commonly, in other rocks. For instance, reef-limestones make up but a small part of the Carboniferous Limestone. At many localities it is composed of muddy limestones with shale partings. Such beds yield numerous solitary zaphrentoid corals. Discoidal solitary corals, such as *Cyclocyathus* (No. 42) can be found in the stiff clays of the Gault, whilst the Eocene clays sometimes yield compound corals such as *Litharea* (No. 39).

4 POLYZOA

The polyzoans are colonial aquatic organisms, mainly marine, though a few live in freshwater. Americans prefer the name **Bryozoa** (*moss animal*) after the habit of certain forms. The colony is built up by many exceedingly minute *zooids*, each living in a tube-like chamber in a calcareous, or in the case of the freshwater forms, a chitinous framework of considerable beauty and complexity. The polyzoan zooid resembles the coelenterate polyp in having a ring of food-gathering tentacles round its mouth, but it clearly belongs to a different phylum, for the soft parts of the zooid are much more highly organized than are those of the polyp.

As the apertures of the chambers inhabited by the zooids are only fractions of a millimetre in diameter, the study of polyzoans is very much a matter for the specialist, using a fairly high power binocular microscope rather than a hand lens. We shall therefore deal only briefly with this phylum.

Polyzoan colonies vary in form from short stout rods, branching encrusting forms, or lace-like sheets to small compact masses growing outwards and upwards from the point of attachment. Polyzoans are first definitely known from rocks of Ordovician age. They are quite important as reef-binders in limestones of Palaeozoic age, especially the lace-like sheets of *Fenestella* (No. 44) in the Carboniferous and Magnesian and Zechstein Limestones. Shaley layers interbedded with

Fig. 14. *Polyzoa. Above – Block diagram of part of a polyzoan colony. Soft parts in red – Hard parts in black. Below Left – A 'stick' polyzoan. Centre – A 'net-like' (fasciculate) polyzoan. Right – A colony with its branches in contact. Apertures are shown in black.*

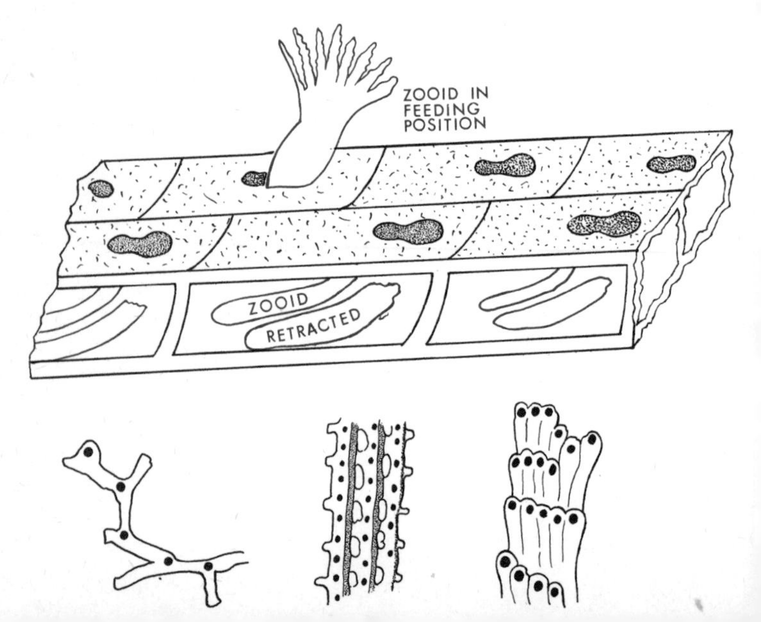

the shallow-water limestones often yield nicely weathered out specimens of 'stick bryozoans' (No. 43), whilst encrusting forms may be found on the shells of brachiopods, etc. Sea urchins from the Chalk also often carry an interesting epifauna including polyzoans. Fossil polyzoans are most common in deposits laid down in shallow clear current-swept waters. They are particularly abundant in the Coralline Crag of East Anglia, for example *Fascicularia* (No. 45). The name of this formation dates from the time when polyzoans were thought to be related to the corals.

5 WORMS AND CONODONTS

Worms are elongate bilaterally symmetrical organisms, capable of movement and with marked differentiation between their two ends; the anterior or head end being usually well equipped with sense organs. The great majority of the worms are entirely soft bodied. Zoologists divide the worms into many phyla but fossil worms are representatives of only one of these, the phylum of the **Annelida** or segmented worms.

At low tide on many modern beaches, one finds abundant evidence of the presence of burrowing worms. Certain worms secrete mucus, which forms a calcareous lining to their burrows. Fossil worm burrows are common fossils at a number of horizons. In parts of the English Upper Greensand and in other sandstones of Upper Jurassic and Cretaceous age the irregularly curved tubes of *Serpulae* (No. 46) may be found in great numbers. The 'pipe-rock' of the west coast of Scotland, now a quartzite, but originally a beach sand of Lower Cambrian age, was named from the numerous worm burrows it contained (No. 47), some of them an inch or more in diameter and many inches in length. These burrows are straight or but slightly curved, but at other horizons, as for example the Lower Lias of the Dorset coast, one may occasionally find U-shaped burrows, similar to those made by the modern lug-worm *Arenicola*. Unquestionable worm burrows grade into the 'trace fossils' (pp. 174–5).

Impressions of the bodies of worms have been found in a few abnormally fossiliferous horizons, such as the Pound Quartzite of late Pre-Cambrian age in South Australia and the Middle Cambrian Burgess Shale of western Canada. These discoveries show that the worms were already well differentiated by late Pre-Cambrian times. Indeed the evidence suggests that the arthropods (p. 163) may have developed from an annelid-like ancestor.

The microfossils obtained from Palaeozoic or later systems may contain minute chitinous, horny or siliceous toothed structures or *Scolecodonts* (Fig. 15). These resemble closely the jaws of living polychaete worms.

A more important group of microfossils are the *Conodonts*. Modern research has shown that many conodonts of distinctive morphology have a limited vertical but a wide geographical distribution and therefore they are of considerable value for zonal purposes. Toothlike or platelike in form they are very minute, less than 2 millimetres in length (Fig. 15). Their systematic

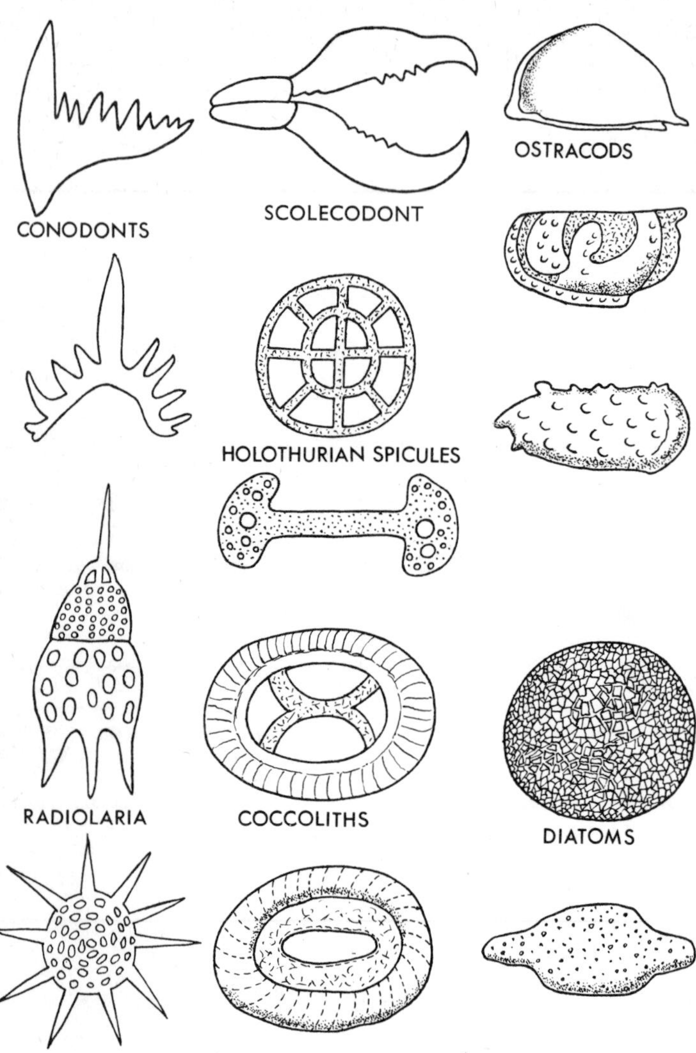

Fig. 15. *Representative microfossils.*

Conodonts	*magnified approximately*	25	*times*
Scolecodont	,, ,,	15	,,
Ostracods	,, ,,	20	,,
Holothurians	,, ,,	30	,,
Radiolaria	,, ,,	300	,,
Coccoliths	,, ,,	1200	,,
Diatoms	,, ,,	400	,,

position is uncertain. They have been grouped with fishes, gastropods, worms and crustaceans. Unlike scolecodonts, conodonts are composed entirely of calcium phosphate. Their wide geographical distribution is another argument against them being parts of bottom dwellers. One possibility is that conodonts may have been the internal supports of structures, such as the gills, in certain fish. They are restricted to rocks of Ordovician to Cretaceous age. Owing to their minute size and the special techniques needed for their extraction, such as prolonged digestion of limestone in acetic acid or the centrifuging of black shale, the study of conodonts is very much a matter for the specialist.

6 BRACHIOPODA

The Brachiopods (*brachia*, arm-*pod*, foot) take their name from a mistake, for their internal *brachia*, used for feeding, were originally assumed to be used for locomotion like the foot of the molluscs. They are marine invertebrates, living mainly in shallow waters, though some have been dredged up from the ocean depths. A few forms can tolerate the brackish water of river estuaries, whilst some live between tidemarks in tropical waters. But today the brachiopods are much less varied and numerous than they were during the geological past, especially in the Palaeozoic Era.

The two main groups of brachiopods are the Inarticulata and the Articulata.

The living *Lingula* is an **inarticulate** form. Its soft parts are enclosed between two thin dark-coloured shells or *valves* composed largely of calcium phosphate, though with some chitin. The two valves, almost identical in shape, have rounded *anterior* and pointed *posterior* ends. The valves gape at the posterior end to allow for the protrusion of a long and strong *pedicle*. Like other brachiopods, *Lingula*, after a brief larval stage, becomes fixed to the sea bottom and remains there for the rest of its life. *Lingula* is attached by its pedicle to the bottom of a burrow in soft mud or silt. The burrow may be as much as 12 inches in depth as compared with the 2-inch, or less, length of the valves. By retracting or extending its pedicle *Lingula* can move up or down its burrow. In the feeding position, the anterior portion of the shell is above the mouth of the burrow. The valves then open slightly so that a current of water can be drawn through the shell by the movement of vibrating cilia on the arms or *brachia*. Short bristles on the margins of the soft parts form three funnels, a central one for the outflowing and one on either side for the inflowing current. As the water flows past the mouth any microscopical food particles are absorbed and the waste products are extruded. The opening and shutting of the valves and the retraction or extension of the pedicle are controlled by a complex pattern of muscles situated close to the *umbones* at the pointed posterior end of the shell. The place of attachment of the muscles are shown by smooth areas or *muscle scars* on the inside of the two valves.

There are a number of unusual features about *Lingula*. In the first place, unlike other brachiopods, it can

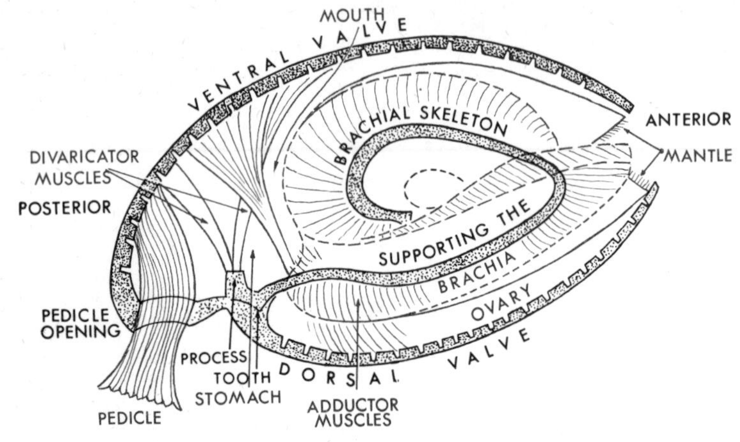

Fig. 16. *Morphology of a terebratulid brachiopod. Soft parts in red – Hard parts in black.*

live in water of abnormal salinity, secondly the living *Lingula* is almost identical with *Lingulella* of the Cambrian rocks. No other macroscopic organism has such a long geological history. Whilst the Cambrian *Lingulella* occurs in rocks which seem to have been deposited under normal marine conditions, the *Lingulas* of the basal beds of the Old Red Sandstone, the *Lingulas* found between the marine bands and the non-marine shales of the Coal Measures, the *Lingulas* of the Deltaic Series of the Jurassic must have lived in brackish water. One can sometimes find specimens of *Lingula* lying at right angles to the bedding planes, obviously in the burrows and sometimes there is a thin pyritic streak beneath the umbones, interpreted as the trace of the pedicle.

A terebratulid such as the living *Magellania* of the coasts of New Zealand is a good example of an **articulate** brachiopod. It shows many differences from *Lingula*. The two valves are strongly convex and unequal in size, but like those of *Lingula*, they are bilaterally symmetrical. The larger valve is the *ventral* or *pedicle* valve with an almost circular pedicle opening or *foramen* in its prominent curved umbo, the smaller valve is the *dorsal* or *brachial* valve. Both valves are composed of opaque calcium carbonate and do not have the dark greenish lustrous appearance of those of *Lingula*. The valves articulate by means of a pair of *teeth* on the ventral valve fitting into a pair of *sockets* on the dorsal valve. Two pairs of muscles, the *divaricators* extend from a projection, the *cardinal process*, between the teeth sockets on the dorsal valve to the ventral valve. Contraction of these muscles causes the posterior parts of the shell to move closer together and the anterior portion to gape. The closing muscles or *adductors* extend from the inside of one valve to the other, so that when they contract the shells close (Fig. 16). In addition there are smaller muscles

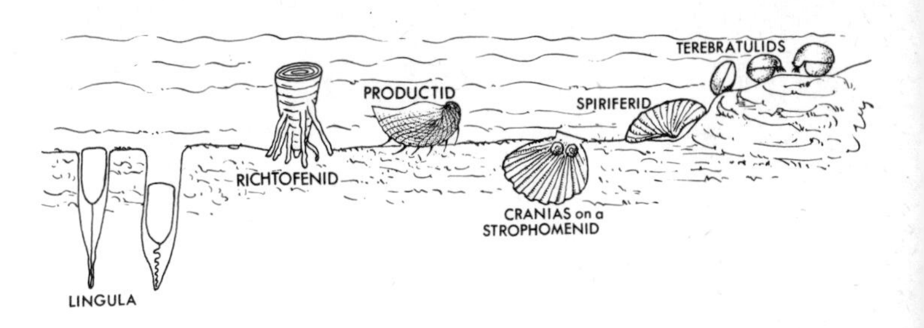

Fig. 17. *Selected brachiopods in their position of life.*

attached to the pedicle. From the hinge line on the dorsal valve, a delicate double loop extends nearly to the anterior margin of the shell and there is a strong median septum. This is the *brachial skeleton* supporting the two brachia of the lophophore, which hang in the mantle cavity. *Magellania* therefore differs from **Lingula** in the unequal size of the two valves, in their calcareous nature, in possessing an articulating mechanism and a brachial skeleton. It lives only in marine waters, usually in colonies, attaching itself to a hard substratum by means of its short pedicle (Fig. 17).

The Geological History of the Brachiopods

The first brachiopods are to be found in rocks of Lower Cambrian age. They are inarticulate forms such as *Lingulella* (No. 48) but at higher Cambrian horizons appear other inarticulate forms with calcareous shells. The articulate brachiopods become common in rocks of Ordovician age. They include a variety of *orthids* (Nos. 49 and 50) with both valves slightly convex and a straight hinge line which is usually shorter than the width of the shell. Externally the valves are ornamented by ridges radiating from the umbones and crossed by concentric growth lines. On each valve between the umbo and the hinge line there is an unornamented *cardinal area.* The orthids of the Ordovician range up to about an inch in width. Internal moulds show the muscle scars and the plates which carry the two large teeth on the ventral valve (see *Nicolella* (*Orthis*) *actoniae*, Fig. 4).

The *strophomenids* are another important group that first appears in the Ordovician rocks. Their hinge line is long and straight at the greatest width of the shells. One valve is convex, the other concave, leaving little space between for the soft parts. Externally they show radial ornamentation, often very closely spaced (No. 4).

The brachiopods of the Silurian rocks are even more numerous and varied and indeed it was at this time that the brachiopods reached their acme. Important groups, in addition to a variety of orthids and strophomenids, are the pentamerids, the spiriferids and the rhynchonellids. The last two groups are *telotrematous*

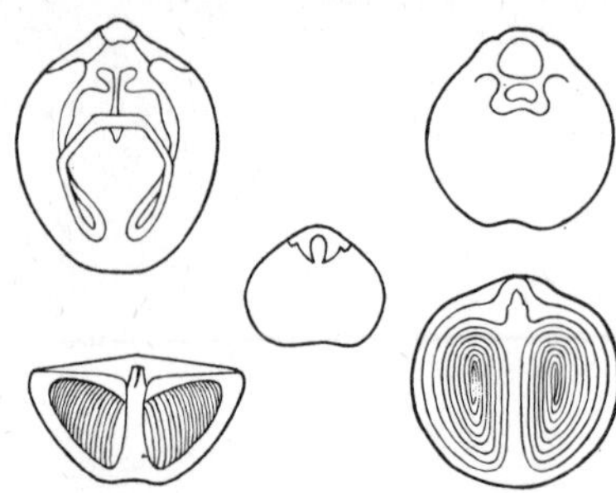

Fig. 18. *Different kinds of brachial skeletons. Top left – Long-looped Terebratulid. Top right – Short-looped Terebratulid. Centre – Rhynchonellid with crurae. Bottom left – Spiriferid with laterally directed spines. Bottom right – Atrypid with spines at right angles to hinge line.*

brachiopods, that is they had brachial skeletons. Both valves of the *pentamerids* are markedly convex, usually with strong ribbing (No. 53), whilst the umbones are prominent, especially the ventral one, and are incurved. The distinctive feature of the pentamerids is the presence, not of a brachial skeleton, but of dental plates uniting to form a long median septum. The older British geologists gave the nickname 'Government Rock' to certain beds of Lower Silurian age in Shropshire. These beds, packed with *Pentamerus oblongus* show the mark of the broad arrow (No. 52), the symbol for Government property. The median septum formed the shaft of the broad arrow, the shell margins its barbs. The *spiriferids* (No. 57) are a distinctive group with well marked cardinal areas on either side of the long straight hinge line. The valves show strong radial ribbing whilst there is a prominent depression or *sulcus* on the ventral with a corresponding *ridge* on the dorsal valve. In anterior view the valves do not meet along a straight or but slightly curved line, as in the groups dealt with above, for there is this marked fold along the median line of the *commisure*. The interior of the shells of spiriferids is largely filled by the brachial skeleton, consisting of two calcareous spiral structures. In most spiriferids the apices of the spires point laterally, but in some they are directed towards the centre of the convex dorsal valve (Fig. 18). The *rhynchonellids* have a triangular outline, the two valves show a prominent radial ribbing, there is a broad and deep ventral sinus with corresponding dorsal fold, so that the line of the commisure is interrupted by a prominent fold. The ventral umbo is sharp, incurved and usually hides the pedicle opening. Internally the brachial skeleton consists only of two short projections or *crurae*.

Brachiopods are very abundant in the sandy and calcareous beds of Ordovician and Silurian age. As already mentioned (p. 20), this shallow water facies is zoned by its trilobite-brachiopod assemblages. At certain localities, as on View Edge in Shropshire, one can find fossil shellbanks packed with specimens of one species of brachiopod, in this case the pentamerid *Conchidium knightii* (No. 53), but this is unusual and at most places one finds brachiopods in considerable variety. Shale partings usually yield nicely preserved specimens in the solid, whilst seams of 'rottenstone' provide good internal moulds. In the highest beds of the Silurian, however, the brachiopod fauna becomes much more restricted. The bedding planes may be crowded with specimens, but usually only two species are represented, the rhynchonellid *Camarotoechia nucula* (No. 51) and *Chonetes striatella*, whose name was recently changed to *Protochonetes ludloviensis*, a small strophomenid, whose pedicle opening became closed in adult life, so it attached itself to the substratum by the development of many small spines along the area of the ventral valve. Specimens can often be found with these spines in position (No. 56). Only *Lingula* was able to tolerate the change from normal marine to brackish and finally freshwater conditions, so we find *Lingula* ranging on into the basal beds of the Lower Old Red Sandstone, well above the horizon at which the rhynchonellids and strophomenids disappear.

The marine Devonian rocks of south-west England, the Ardennes and the Rhineland, yield a large variety of brachiopods, which are used for zonal purposes. Spiriferids are particularly prominent and they are joined by the last order of the telotrematous brachiopods, the *terebratulids*. The main features of the terebratulids are similar to those already given for *Magellania*. Two smooth minutely punctate convex valves with a prominent pedicle opening in the ventral umbo (No. 59), a commissure that is straight or gently curving and internally a brachial skeleton in the form of loops. The pentamerids are much less important than in the Silurian, but orthids and strophomenids persist.

In the limestone facies of the Carboniferous brachiopods continue to be important. *Productus* and *Pugnax* are two particularly characteristic Carboniferous forms. *Productus* (No. 54) a strophomenid, had a concave dorsal and a strongly convex ventral valve with strongly incurved umbones. Externally the shell had a strong radial ornamentation crossed by well-marked growth lines. The width of the shell was often increased by 'ears'. The gigantoproductids (No. 55) are the largest of brachiopods, ranging up to nearly 2 feet in width, as compared with the few inches or less of the majority of brachiopods. In its living position, *Productus* was not fixed by a pedicle, but rested on its ventral valve, in a position rather similar to that of the Mesozoic oyster *Gryphaea* (No. 92). The anchorage of many productids was improved by the development of spines on the ventral valve. In most fossil specimens these have been broken off and only their bases can be seen. But sometimes one can find productids, which have been replaced by silica. Then the matrix can be dissolved gradually away with acid, leaving the spines, which may be as long as the rest of the shell, in their position in life. *Pugnax* is an almost smooth rhynchonellid. The median fold and corresponding sulcus are so greatly developed that *Pugnax* has almost the shape of a tetrahedron (No. 58). The umbo, though not too prominent, enables one to determine which is the ventral and which is the dorsal valve.

The lower beds of the Zechstein and Magnesian limestones of Permian age contain rhynchonellids and productids, usually strongly spinose, but they died out upwards, as the salinity content of the sea water increased. In Permian rocks of the Tethyan belt in Asia and in Texas are to be found the *richtofenids*, an aberrant group of brachiopods, in which the ventral valve developed into a cone, whilst the dorsal valve became small and lid-like (Fig. 17). In their assumption of a coralloid-form, the richtofenids resemble certain aberrant sessile lamellibranchs (p. 149).

By the close of the Palaeozoic Era, the strophomenids, the orthids and the pentamerids had passed into extinction, whilst the spiriferids only lasted into the early part of the Jurassic Period. The inarticulate brachiopods persisted, but they were never important, except in the special environments dominated by *Lingula*. In the Jurassic and Cretaceous rocks, especially in the limestones, a great variety of rhynchonellids and terebratulids are to be found. The rhynchonellids include the spinose acanthothyrids; the terebratulids, forms like the aptly named sphaeroidothyrids (No. 61), others with greatly elongated ventral umbones or the rather bag-shaped digonellids (No. 60). Many horizons in the Jurassic are recognizable owing to their highly distinctive assemblage of brachiopods, whilst the zones of the Middle Chalk are named after brachiopods. The Danian Chalk yields *Crania* (No. 63), an almost square inarticulate form, which was cemented to the sea floor by its ventral valve.

Brachiopods, however, play but an insignificant part in the faunas of the marine Tertiary beds. They occur mainly in the rocks laid down in clear water, such as the Crags of East Anglia, which yield *Terebratula maxima*, up to 4 inches in length, and as its specific name implies a 'giant' amongst terebratulids. In Recent seas, as already stated, brachiopods are widely distributed, but never a major part of the marine fauna, as they were in Palaeozoic times.

The efficient hinge apparatus of the Articulata, particularly of the Telotremata, usually ensures that the two valves do not become separated after death. This is different from the lamellibranchs, whose valves were joined by an organic ligament (p. 148), which was prone to decay and then the two valves separated. The detailed classification of the telotrematous brachiopods is based very largely on the precise nature of the brachial skeleton (Fig. 18). This is hidden in the normal 'solid' brachiopod. Sometimes one does find separate dorsal valves and if so, it may be possible to remove the matrix and see if the brachial skeleton is preserved in its entirety. But these are not common and it is usually a matter of carefully grinding down a solid brachiopod, stopping at every fraction of a millimetre, to draw or photograph the cross-section of the brachial skeleton. Then from such a series of closely spaced drawings, the brachial skeleton can be reconstructed in the solid, an extremely laborious process, clearly only a matter for the brachiopod specialist.

7 MOLLUSCA

The Mollusca (*soft bodied*) comprise an extremely varied phylum, divided into the following five classes:

(*a*) Amphineura (chitons).
(*b*) Scaphopoda (tusk shells).
(*c*) Cephalopoda (squids, octopods, etc.).
(*d*) Lamellibranchiata or Pelecypoda (clams).
(*e*) Gastropoda (snails).

The first three classes are entirely marine, as are the majority of the other two classes, but certain of the aquatic lamellibranchs inhabit either brackish or freshwater, whilst the gastropods include both aquatic and terrestrial forms. Molluscs vary in size from minute forms to the deep sea squids, the largest of all invertebrates, 50 feet or more in length, whilst the shells of some of the clams living in tropical waters may weigh as much as 600 lb.

The distinctive features of the Mollusca are complete or nearly complete absence of segmentation; the body is normally elongate and bilaterally symmetrical; in all classes except the lamellibranchs, there is a distinct head with sensory organs; the viscera are enclosed by the body wall, the lower part of which is modified to form the *foot* used for locomotion, whilst the upper part (*the mantle*) secretes a shell composed of calcium carbonate. In the chitons the shell is segmented and formed of eight pieces, the lamellibranchs are bivalves, whilst the other classes are univalves. In certain of the cephalopods the shell is internal and not, as in the other classes, external.

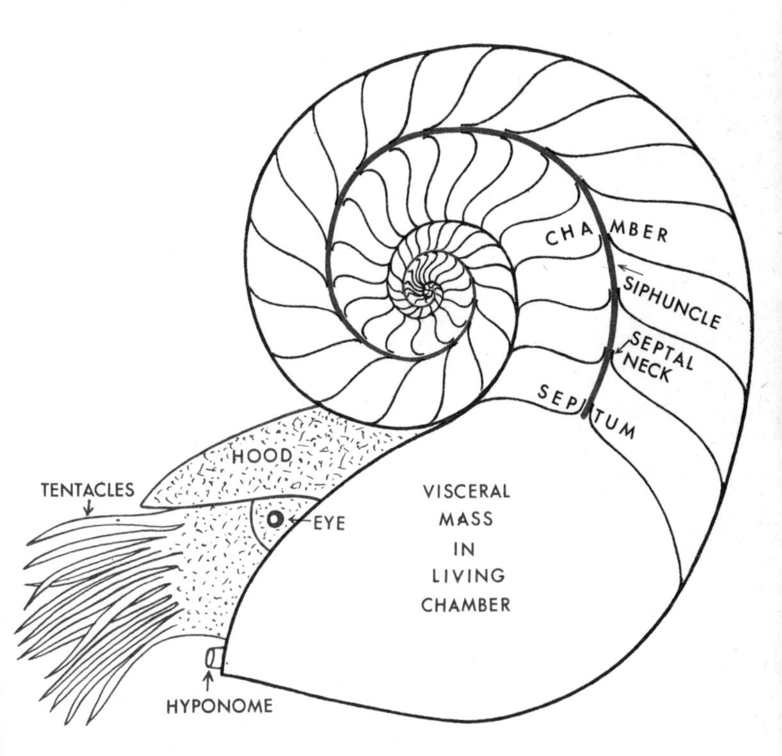

Fig. 19. *Morphology of a nautilus. Soft parts in red – Hard parts in black.*

7a and *b* Amphineura and Scaphopoda

Although both the chitons and the scaphopods range from the Lower Palaeozoic to the Recent, they are neither important nor common as fossils. At some horizons in the marine Lower Tertiary rocks, the delicate tapering tubes of the scaphopod *Dentalium* (No. 114) are fairly abundant. The scaphopods lived partly embedded in the mud or sand of the sea bottom with the smaller end of the shell protruding. They were deposit feeders, abstracting organisms such as Foraminifera from the mud or sand. The chitons crawl sluggishly over the sea floor, feeding on algae, hydroids and other benthos.

7c Cephalopoda

Modern cephalopods are the squids, octopods and **Nautilus**, but in the past, particularly in the Mesozoic, the cephalopods were much more varied. They are a highly organized group, exclusively marine, mobile and catching their prey with prehensile tentacles, often armed with suckers. The tentacles surrounding the mouth on the definite head are a modification of part of the molluscan foot, hence the name cephalopoda (*cephalo*, head – *poda*, foot). Modern cephalopods are sub-divided by their number of gills into the *Dibranchiata* (squids, cuttlefish, octopods) and the *Tetrabranchiata* (*Nautilus*). But unfortunately gills have not been preserved in the extinct cephalopods and therefore, whilst the classification of cephalopods given below is hallowed by long usage, it may not be scientifically correct.

Tetrabranchiata

The modern pearly *Nautilus* is very important, for it is the sole modern representative of the Tetrabranchiata. Its living area is the south-western Pacific between the islands of Fiji and the Straits of Malacca, but after death its buoyant shells may be carried far beyond these limits. The aragonitic shell is coiled in a plane spiral with the outer *whorls* almost completely covering the inner whorls, so that there is but a small *umbilicus* or space between the inner side of a whorl. The outer surface of the shell is smooth, apart from faintly marked growth lines. Internally the shell is divided into as many as thirty-six chambers by *septa*, which meet the inside of the outer wall along a slightly curving line known as the *suture* (Fig. 19). The soft parts are contained in the living chamber between the outermost septum and the aperture of the shell. *Nautilus* has a well defined head with a pair of simple eyes and about ninety tentacles, without suckers, round the mouth, which is armed with both a pair of horny jaws and a radula. Above the head is a tough hood, which closes the aperture when

Cephalopoda	Tetrabranchiata	Nautiloidea	Cambrian–Recent
		Ammonoidea	Devonian–Cretaceous
	Dibranchiata	Belemnoidea	Carboniferous–Cretaceous
		Sepioida	Jurassic–Recent
		Teuthoida	Jurassic–Recent
		Octopoida	Cretaceous–Recent

the creature withdraws completely into the living chamber. Beneath the head and pointing forwards is a funnel or *hyponome*, from which water can be ejected with considerable force. The septa are concave backwards. The centre of each septum is perforated by backward pointing septal necks. Through these extends a fleshy tube or *siphuncle*. *Nautilus* is a nektonic form. The specific gravity of the complete organism (shell, soft parts and gas-filled chambers) is very close to that of the water in which it lives. It can probably alter its specific gravity slightly, and hence its depth of flotation, by adjustment of the amount of gas- and liquid-filled space in the last few chambers. The internal gas pressure within the shell is slightly less than one atmosphere. The shell must be very strong for *Nautilus* is known to live at depths of at least 200 metres, where the hydrostatic pressure on the shell reaches nearly 20 atmospheres. The shell is also very stable in its normal swimming position (see p. 144). It is a good swimmer, jets of water ejected from its *hyponome* causing it to move quickly upwards and backwards

The Geological History of the Tetrabranchiate Cephalopods

The earliest fossil nautiloids are found in rocks of Cambrian age. The Ordovician rocks of the Baltic coasts and parts of the U.S.A. have yielded a very considerable variety of nautiloids. The majority are straight forms and some such as *Endoceras* are of exceptional size, specimens up to 13 feet in length having been found. Their walls are thick and there is a great development of calcareous material along the siphuncle. *Orthoceras* (No. 64), a straight nautiloid with straight suture lines, is to be found at many localities in the Palaeozoic rocks. Other Palaeozoic nautiloids had curved shells, e.g. *Cyrtoceras*, in others the shell was coiled with each whorl slightly enveloping the inner whorl, e.g. *Ophioceras*, whilst *Lituites* (Fig. 21) is an abnormal form with the first formed part of the shell strongly coiled, the later parts straight. In the Upper Palaeozoic rocks nautiloids are less prominent; the majority of them are coiled, in one genus, *Lorieroceras*, in a helicoid spiral, and the shells of some genera are ornamented by spines or ribs, instead of being smooth as is the case with most other nautiloids. The decline of the nautiloids continued throughout the Mesozoic and Tertiary. Those forms that are found (No. 65) are increasingly involute (tightly coiled) and differ but slightly from the modern *Nautilus*.

In Devonian times the **ammonoids** developed from a nautiloid ancestor, possibly the straight *Bactrites* with its siphuncle located near the ventral margin. Unlike the slowly evolving nautiloids, the ammonoids were a most varied and geologically important group. They can be conveniently divided on the characters of the suture line, into the goniatites (Devonian to Permian), the ceratites (Devonian to Triassic) and the ammonites (Jurassic and Cretaceous). The distinctive features of the *goniatites* are first that the suture lines are not straight or slightly flexuous as in the nautiloids, but are sharply angular (No. 66), the forward projections being *saddles*, the backward ones *lobes*, secondly the siphon instead of being central is on the outer (ventral) side of the whorl, whilst the septal necks point both backwards and forwards. Goniatites are smaller than the majority of the nautiloids, rarely exceeding six inches in diameter, they are tightly coiled and show considerable variation in whorl section. Ornamentation is another feature that is much more developed in the goniatites. Some, e.g. *Manticoceras*, bear closely spaced flexuous ribs, *Reticuloceras* takes its name from the network-like pattern of longitudinal and transverse lines, others such as *Gastrioceras* have short tubercles carried on stout ribs near the inner (dorsal) margins of the whorls. The smooth genera can be distinguished by the variations in the form of the suture lines. These are clearly shown wherever the thin outer shell has been worn away (No. 66). Individual genera have a short time-range, a wide geographical distribution and are therefore excellent zone fossils. Unfortunately the goniatites are only found commonly in the shales of Upper Palaeozoic age. In the limestone facies, particularly in the more massive limestones, they are very much rarer and therefore these beds have to be zoned on a broader framework using assemblages of the benthonic corals and brachiopods instead of the pelagic goniatites. Goniatites are of great

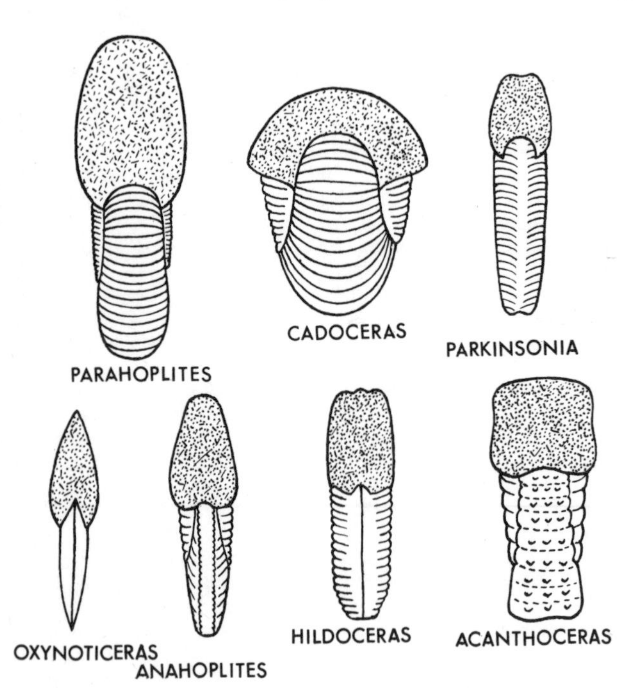

Fig. 20. *Selected ammonites to show range of variation in coiling, shape of venter and ornamentation.*

importance in zoning the Devonian rocks of the Rhenish Schiefergebirge, but unfortunately they are not nearly so abundant in the Devonian rocks of south-west England. The shales of the Millstone Grit have been subdivided in great detail by use of goniatites. They occur either as impressions crushed flat on the bedding planes or they may be preserved in the solid in calcareous nodules or in thin seams of limestone. Certain thin limestones in western Eire are packed full of beautifully preserved 'solid' goniatites. They also occur in the marine bands of the lower part of the Coal Measures.

In the Upper Devonian strata there also appear the ancestors of the ceratitoid ammonoids, with suture lines showing rounded saddles and frilled lobes. The marine Permian and Triassic rocks of the Tethys have yielded a great number of ceratitoid ammonoids varying greatly in ornamentation, whorl shape and tightness of coiling. They even include some genera which are uncoiled and others coiled in a helicoid spiral. *Ceratites* (No. 67) is a common fossil of the German Muschelkalk. By the close of the Triassic Period, all the host of ceratitoids had died out, except for one family the phylloceratids, a stable stock, strongly involute with their thin shells covered with fine growth lines (No. 69). From this family developed the ammonites. The marine Jurassic and Cretaceous rocks of the British Isles, France, Germany, Denmark and extreme southern Sweden are zoned in great detail by their wealth of ammonites, all characterized by extremely complex sutures with the lobes and the saddles elaborately frilled.

It is impossible in the space available to illustrate or even refer to more than a few of the commoner ammonites. Those mentioned are selected to indicate the limits of variation within the ammonites. In cross-section from the smooth, tightly coiled *Harpoceras* (U. Lias) (No. 71) to the broad *Cadoceras* (Middle Jurassic) (No. 72), in ornamentation from smooth forms to strongly ribbed forms like *Asteroceras* (Lias) (No. 1) or *Perisphinctes* (U. Jurassic) or those with both ribs and spines such as *Douvilleiceras* (No. 74) (Gault). The venter may be sharply angular as in *Oxynoticeras*, rounded as in *Perisphinctes*, with a prominent *keel* as in *Hildoceras* (Lias) or with a groove or *sulcus* (Fig. 20) as in the hoplitids (Cretaceous) (Nos. 5 and 75). Few of the Lower or Middle Jurassic ammonites exceed 9 inches in diameter, but the Upper Jurassic and Lower Cretaceous rocks yield a number of much larger forms, such as the perisphinctids of the Portland Stone of England, often 2 feet in diameter, whilst the German Cretaceous has yielded *Pachydiscus*, 6½ feet in diameter, the largest known ammonite. Another feature of the Cretaceous ammonite fauna is the presence of many *heteromorphic* forms, showing varying degrees of uncoiling, e.g. *Scaphites* (No. 76), *Hamites* (No. 77), or of coiling in a helicoid spiral, e.g. *Turrilites* (Fig. 21). The marine Cretaceous rocks up to the level of the basal beds of the Chalk yield ammonites in great abundance and variety, but above that horizon the ammonites are much rarer and none survive into beds of Tertiary age. The last known ammonite *Baculites* is a straight form, though with the complex ammonitic suture (No. 78).

In many Middle and Upper Jurassic ammonites, the margin of the aperture, instead of being smooth as in *Nautilus*, the goniatites and the ceratitoids, is narrowed by two forwardly projecting *lappets* (No. 73) and less commonly by a long curving *rostrum* on the venter. It has long been noted that the same bed can yield ammonites identical in ornamentation, type of coiling, etc., except that the larger forms have smooth apertures, whilst the smaller forms show lappets and sometimes a rostrum. The significance of this *dimorphism* is not fully understood, but some specialists regard it as sexual, the larger forms being females, with the lappets and rostrum forming a brood chamber. Certainly in modern *Nautilus* the male shell is slightly larger than that of the female, but the difference between the sexes is not nearly so great as in these ammonites assigned in the past to different species and even genera, but which many would now pair as male and female members of the one species.

When an ammonoid withdrew into its shell, it could close the aperture by either a pair of calcitic plates, *aptychi*, or by a single horny *anaptychus* plate. Aptychi are only found in Mesozoic rocks, anaptychi, which are much rarer, in beds of Devonian to Cretaceous age. Aptychi with their straight hinge line, slight convexity and delicate ornamentation bear some resemblance to lamellibranch shells (No. 79), but sufficient have been found in position, closing the apertures of ammonites, to leave no doubt as to their function. In the Mesozoic rocks of the Alps there are beds yielding numerous aptychi, but the shells of ammonites are either extremely rare or are completely absent. This may be due to current sorting or alternatively the aragonitic shells of the ammonites have all disappeared owing to solution or some other diagenetic change, whilst the calcitic aptychi have survived.

Mode of Life

The mode of life of ammonites is still a matter of conjecture. In *Nautilus* the shell seems to have functioned both for protection and also for buoyancy. The late Sir Arthur Trueman calculated the positions of the centres of gravity and the centres of buoyancy for a number of genera. In evolute many whorled shells these are very close together, but the involute forms must have been more stable with the aperture facing upwards as in *Nautilus*. The two centres were still further apart in the heteromorphs, so crawling over the bottom on their tentacles must have been almost impossible. Those ammonites with an oxycone shell must have been good swimmers, others were probably much less effective swimmers and may have been bottom dwellers. What they fed on is quite uncertain.

Dibranchiata

Belemnites, the so-called 'devil's thunderbolts', are common fossils in the marine clays of Jurassic and Cretaceous age. All that is usually found is the strong pointed cigar-shaped *guard*. When broken the guard is seen to be built up of radiating fibres of calcite, sometimes showing slight concentric colour banding similar to the growth lines on sections across tree trunks. At the anterior (non-pointed) end of the guard, there is a deep conical depression or *alveolus*. In better preserved specimens, the alveolus is infilled by a chambered calcareous *phragmocone* (No. 83) from one side of which projects forward the *pro-ostracum*. Very exceptionally specimens have been found showing impressions of the soft parts. The guard, the

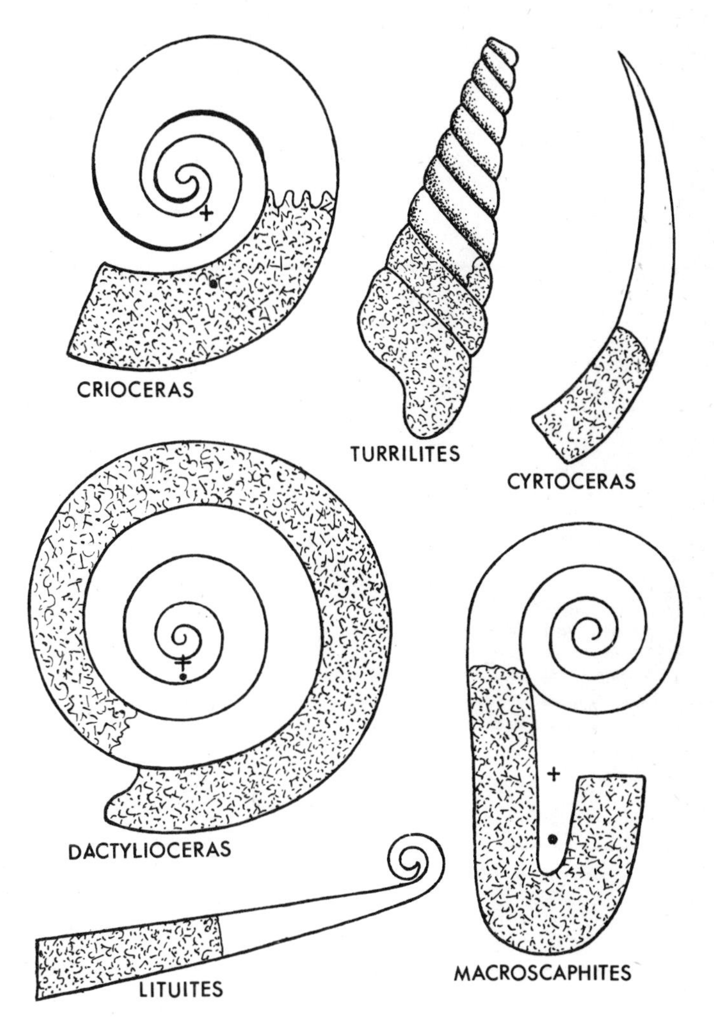

Fig. 21. *Probable living position of selected cephalopods. Circle – Centre of gravity. Cross – Centre of buoyancy.*

phragmocone and the pro-ostracum formed the internal skeleton of a creature resembling the modern cuttlefish, with streamlined body, a head with a beak and prominent eyes and ten arms (Fig. 22). The arms bore suckers for seizing prey, but unlike those of the modern cuttlefish and octopods, the shorter arms were also furnished with horny hooks. On one side of the head was a hyponomic funnel, whilst exceptionally the ink-sac, from which a protective cloud could be ejected into the water has been found. The ink of some Jurassic belemnites has been so well preserved that it could be dissolved out and used. Like the modern cuttlefish the belemnites must have been nektonic carnivorous creatures, probably gregarious, judging from the abundance of their guards at certain horizons. They could use either their arms for locomotion or when need be, could dart rapidly backwards by ejecting water from their funnel.

The earliest belemnoids of Carboniferous, and possibly Devonian, age had straight slender phragmocones projecting far beyond the guard. It is probable that they had developed from a nautiloid-like ancestor. By Jurassic times, however, the phragmocone of the majority of belemnites had become much smaller proportionately and did not extend beyond the alveolus. There is considerable variation in the shape of the guards of the Jurassic and Cretaceous belemnites. In the *lanceolate* forms (No. 82), the guards are of constant diameter apart from the conical posterior ends, the maximum diameter of the tapering *hastate* forms (No. 81) is near the posterior end, whilst *mucronate* forms (No. 80) have a small bulb at the posterior tip, though this is easily broken off. Other features of classificatory value are the shape of the guards in cross-section, the depth and angle between the sides of the alveolus and the presence or absence of longitudinal grooves, best developed towards the anterior end. By features such as these, a large number of different genera of belemnites have been recognized. It has been shown at numerous horizons that belemnites

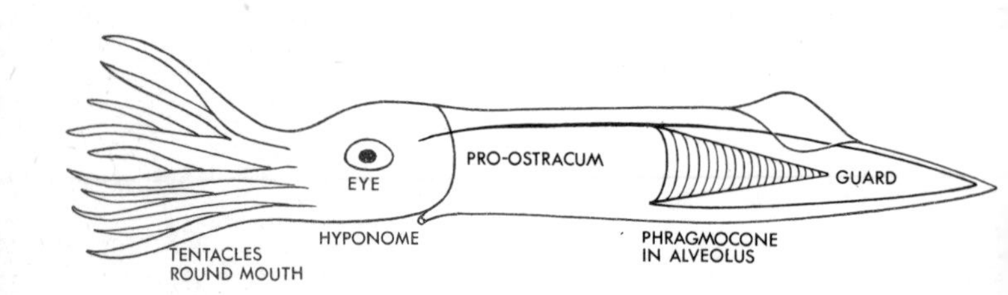

Fig. 22. *Reconstruction of a belemnite. Soft parts in red – Hard parts in black.*

Fig. 23. Right. *Morphology of a lamellibranch. Soft parts in red – Hard parts in black. N.B. This is a generalized Active form.*

can be used for zonal purposes, but usually the same horizons also contain ammonites which are more readily identifiable. However, the belemnites have a longer time range than the ammonites and the highest beds of the Chalk are zoned by belemnites such as species of *Belemnitella* (No. 80). The majority of the belemnoids passed into extinction by the close of the Cretaceous Period, but there are a very few recorded from rocks of Eocene age. Belemnites are often found as derived fossils in the boulder clays, fluvio-glacial outwash gravels and river gravels of Pleistocene age, for their stout guards were able to withstand a great deal of abrasion.

Of the modern dibranchiate cephalopods, little is known of the geological history of the soft bodied *octopoids*, though an impression of one has been found in the Upper Cretaceous rocks of the Lebanon. The *teuthoids*, the true squids, some of which are of gigantic size, probably developed in early Mesozoic times as an offshoot from the belemnites. They have a horny 'pen', a rudimentary internal skeleton, perhaps analogous to the pro-ostracum of belemnoids. The remains of such 'pens' have been found very rarely in Jurassic and Cretaceous rocks. The fossil record of the *sepioids* is more complete. The modern *Sepia*, the cuttlefish, can be compared to a dorso-ventrally flattened squid. Like the octopus it is primarily a bottom dweller, lying in ambush for its prey. Its shell, the cuttle bone, which is sometimes washed up on our beaches, is aragonitic being composed of numerous septa joined by short pillars. One can trace in the Tertiary rocks a series of forms showing the gradual disappearance of the guard and the increase in size and flattening of the phragmocone leading up to the modern *Sepia*. A modern deep-sea, free-drifting cephalopod is *Spirula*, whose skeleton is an internal coiled and chambered shell. A sequence of forms is known from the Tertiary rocks, such as *Belosepia* (Eocene) and *Spirulirostra* (Miocene), showing the disappearance of the guard and, in this case, the coiling of the phragmocone.

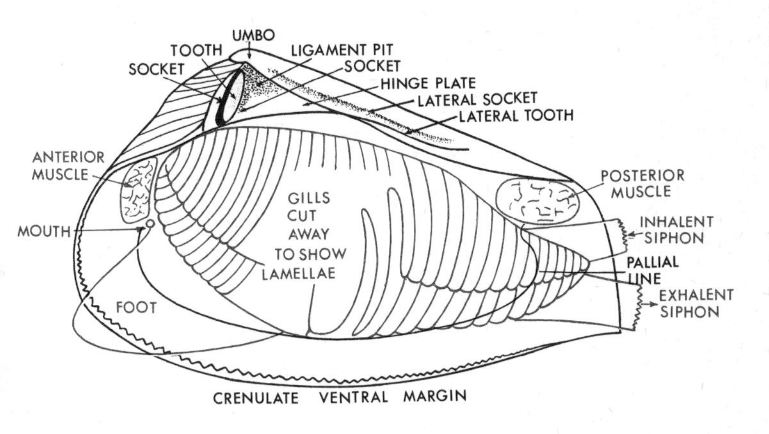

7*d* Lamellibranchiata

The members of this important class are usually called **Lamellibranchs** in Britain, after the lamellar form of their gills. Americans prefer to use either the scientific term **Pelecypoda** (*hatchet foot*) or the popular name **clams**. The term **Bivalvia** is also used, but this may cause confusion for the brachiopods are also bivalves. One obvious distinction is that the two valves of the brachiopod shell are not equal in size as is the case in the majority of the bivalved lamellibranchs.

Mode of Life

We describe first as a typical lamellibranch, *Crassatella sulcata* (No. 84), a common fossil of the Upper Eocene beds of the Hampshire Basin. The soft parts are enclosed between two almost identical *valves*, made mainly of calcite, with an outer prismatic and an inner lamellar layer, whilst the inside of the valves has a pearly appearance due to a thin layer of aragonite. The outside of the valves bears strong concentric ridges. The valves are not symmetrical about a line at right angles to their surface. The shorter end is the *anterior* and hence it is easy to decide whether a single valve is a left- or a right-hand one. There is a well marked *umbo*, which is dorsal and pointing forwards. The interior ventral margin is crenulate. Parallel to it is a well-marked groove, the *pallial line*, marking the inner margin of the thickening of the mantle where new shell is secreted. The pallial line terminates both anteriorly and posteriorly at two distinctive areas, to which the *adductor muscles* were attached. Beneath the umbo is a well-developed *hinge plate*, bearing in the right-hand valve one prominent and one small tooth and in the left-hand valve two prominent teeth, fitting into corresponding sockets on the other valve. Behind the umbo there is a well-marked pit for the rubbery-like *ligament*.

Crassatella is an *active lamellibranch* crawling over the sea floor on its foot and feeding by protruding its siphons. A current of water is drawn in through the dorsal siphon, oxygen is absorbed by the blood circulating within the gills and any microscopic food particles are conveyed by ciliary movement to the mouth. The waste water is then extruded through the ventral siphon (Fig. 23). To feed or to move the lamellibranch must open its shell. If the adductor muscles are relaxed, the contraction of the ligament will cause the two valves to pivot on the hinge formed by the teeth and sockets and gape ventrally. Contraction of the adductor muscles causes the shell to close.

Some lamellibranchs, notably the scallops (*Pecten*) (Fig. 24) and the file shell (*Lima*) are able to move quite rapidly through the water by clapping the valves together and ejecting water suddenly from the gill cavity. The pectens normally rest on the sea bottom, lying on the right valve so that the shells are horizontal, instead of being held vertically as in crawling forms, such as *Crassetella*. The pecten shell shows other changes (No. 86). It is *monomyarian*, not dimyarian, with only one large adductor muscle. The shells are light and corrugated. In many pectens they are almost bilaterally symmetrical, being only slightly less rounded at the anterior end. Teeth are very poorly developed, whilst there is a distinct 'ear' on either side of the umbo. On the anterior side of the right valve, there is a well marked notch for the

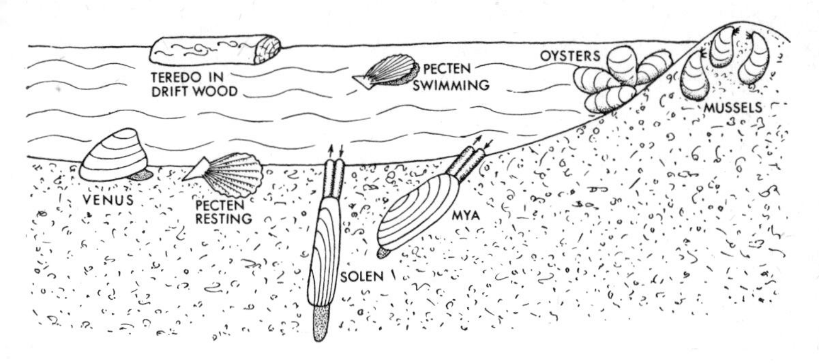

Fig. 24. *Selected lamellibranchs in their position of life.*

passage of the *byssus*, fibres by which the shell can be attached to the sea floor (No. 88).

The mussels (Fig. 24) also attach themselves by a strong byssus. Their valves are roughly triangular in shape, widening obliquely away from the prominent umbones (No. 89). The modern *Mytilus* is extremely gregarious, forming densely packed colonies in shallow water. It can withstand great variations in salinity, and by tightly closing its shell so as to retain water it can stand exposure to the air between tides. *Gervillia* and *Inoceramus* are well-known fossil lamellibranchs which attached themselves by a byssus.

Other *sessile lamellibranchs* cement themselves by the shell to the sea floor. In such forms the two valves are unequal in size, the dentition is much modified, whilst there is only one adductor muscle. The oyster family rest on the left valve, so it is the right valve that is reduced in size. In *Ostrea* (No. 91) the two valves diverge from the umbones, in *Gryphaea* (No. 92) the left valve is thickened and the umbones are incurved, whilst in *Exogyra* (No. 90) the umbones are twisted sideways and in *Lopha* (*Alectryonia*) (No. 93), the shell is strongly plicated. *Ostrea* can tolerate brackish water, but the other members of the family seem to have been restricted to water of normal salinity. All these members of the family of the Ostreacea are abundant in the Mesozoic and Tertiary rocks.

The Upper Jurassic and Cretaceous limestones of southern Europe contain abundant *rudists*, the most aberrant of the lamellibranchs. These are sessile forms of coralloid habit, with the right valve greatly developed and the left valve small and lid-like, but with thick specialized teeth on its inner side. In some of the rudists, e.g. *Toucasia*, the right valve is twisted into a spiral, but in the hippuritids it is in the form of a straight thick-walled cone with longitudinal ribbing. Such rudists may be up to 3 feet in length, with their thick walls penetrated by an intricate system of canals. These forms were reef dwellers in the warm clear waters of the margins of the Tethys. Rudists are not uncommon in the Cretaceous limestones of the south-western side of the Paris Basin, but further north there are only a few records of them from the Cretaceous rocks of north-east France and Britain.

The other major modification shown by the lamellibranchs was for a *burrowing mode of life*. Many of the active forms, when not crawling over the sea floor, buried themselves to a shell-length or more in the soft deposits (Fig. 24). The shells of such forms do not show much modification, except for the development of a notch or *sinus* in the pallial line, due to the size and length of the siphons. In the real burrowers, such as *Pholadomya*, the anterior end of the shell is markedly shortened, whilst the valves are not in contact posteriorly, but show a permanent gape to allow for the passage of the thick and long siphons (No. 94). The internal ligament is often borne on a prominent spoon-shaped structure, the *chondrophore*. The razor shells, the *solens*, show further adaptation, for their valves are very long and thin and do not close either anteriorly or posteriorly to allow for the protrusion of the large foot and siphons (Fig. 24). Razor shells can bury themselves in soft mud or silt with surprising speed. Other burrowers, such as *Lithophagus* (the date mussel) can burrow into soft rock. The date mussels attack limestones by secreting acid from special glands. Their own shell is protected by a thick brown, horny outer layer or *periostracum*. The shells of the piddock, *Pholas*, bear many teeth-like projections and these rasp against soft rock as the valves are rotated around the powerful foot. Finally *Teredo*, the 'ship worm', which is a lamellibranch. After a *Teredo* larva settles on wood it will develop into the adult worm-like form. The only hard parts are the two small valves used for driving the burrow along the grain of the wood. As the burrow lengthens the soft body elongates, for the siphons remain at the opening of the burrow. Specimens of fossil wood can often be found in clays of Mesozoic and Tertiary age with the wood riddled by burrows similar to those of the modern **Teredo.** The walls of the burrows usually showing a thin calcareous lining due to the secretion of lime-charged mucus (No. 96).

Geological History of the Lamellibranchs

In the Lower Palaeozoic rocks lamellibranchs occur but infrequently, especially in beds laid down under marine conditions. Those found, such as the members of the *Modiolopsis* group (No. 97) show primitive characteristics, for they are thin shelled, the dentition is but poorly developed, whilst the musculature for opening and closing the shells must have been complex, judging by the pattern of the muscle impressions. Certain lamellibranchs were able to survive in brackish and even freshwater. In the highest Silurian and earliest Devonian beds, laid down on the margins of the northern landmass (p. 113), one can trace the upward passage from marine beds yielding an abundant and varied assemblage of brachiopods into strata in which the brachiopods die out but lamellibranchs, gastropods and ostracods (p. 136) become more common. At higher horizons, as conditions became more adverse, even these tolerant forms were unable to survive. In the Upper Old Red Sandstone the large freshwater mussel *Archanodon* is not uncommon. Lamellibranchs are to be found, though not too commonly, in the marine limestones of Carboniferous age. The marine shales, especially those of Upper Carboniferous age, yield an assemblage of goniatites (p. 143) and thin-shelled pectiniform lamellibranchs, such as *Dunbarella* (No. 98). The non-marine shales of the Coal Measure rhythm (p. 113) contain 'mussel bands' crowded with specimens of *Carbonicola* (No. 99), *Naiadites*, etc. These freshwater mussels are very important, first for the zonal subdivision of the Coal Measures and secondly for providing material for the study of fossil communities by statistical methods. It is only at a limited number of horizons that one can find a true fossil community. Then one can study the range of variation of its members and attempt to relate slight changes in shell form to ecological conditions as shown by minor but significant changes in the grain-size or chemical composition of the enclosing sediments.

In the Mesozoic rocks the lamellibranchs are much more numerous and varied. The clear water deposits, the limestones and the sandstones, yield many pectens, some such as *Entolium* (No. 86) of the Upper Greensand with strong radial ribbing, others such as *Neithea* (No. 87) with concentric ribbing, together with numerous *Phodadomyas* (No. 94) and related burrowing forms. *Trigonias* with backwardly pointing umbones, grooved teeth (*schizodont dentition*) and the ornamentation different on the anterior and posterior parts of the shell, are common fossils in the Jurassic rocks (No. 102), whilst *Inoceramus* (No. 104) is to be found throughout the Chalk. Broken specimens of *Inoceramus* can easily be recognized by the abnormal development of the prismatic layer of shell in this genus. The closely spaced prisms of calcite are at right angles to the shell walls. Species of *Inoceramus* are used for zonal purposes in the Chalk, particularly that of north Germany, but it is necessary to find fairly complete specimens showing both the shell form and also the details of the concentric ornamentation. We have already mentioned (p. 149) the abnormal sessile lamellibranchs of the Tethyan belt. The more clayey rocks of Mesozoic age yield a rather different assemblage, many oysters (Nos. 90–93), thin-

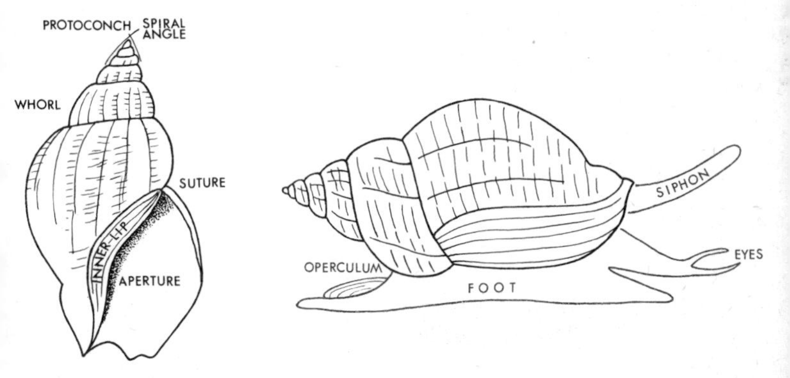

Fig. 25. *Morphology of a gastropod (Buccinum, the whelk). Soft parts in red – Hard parts in black.*

shelled lamellibranchs with a pronounced byssal notch such as *Pteria* (No. 88), or active lamellibranchs such as *Astarte* (No. 107) and *Nucula* (No. 105). Pauses in the deposition of the limestones are marked by surfaces (former sea floors) encrusted by oysters (No. 103) and with numerous borings of various diameters, some of which were certainly made by lamellibranchs. Brackish and freshwater horizons such as the Wealden Beds yield species of *Unio* (No. 100) and *Cyrena*, but whilst these may occur in great numbers, their preservation is all too often poor.

Lamellibranchs and gastropods are very abundant in the rocks of Tertiary age. Those beds which do not yield Foraminifera are correlated mainly by their molluscan fauna. Evolutionary changes amongst the benthonic lamellibranchs proceeded only slowly during the passage of the nearly 70 million years of post-Cretaceous times, so that the correlation of the Tertiary beds is more on a basis of assemblages rather than the vertical range of particular genera or species as is the case with the pelagic ammonites of the Mesozoic. Many of the species of molluscs found in the Tertiary rocks, particularly in the younger Tertiary rocks, have living representatives, so that it is often possible to state with confidence the conditions under which a certain group of beds was deposited, i.e., shallow water marine, brackish or freshwater, under sub-tropical or cold climatic conditions, etc.

The active lamellibranchs of the Tertiary rocks include many pectens, species of *Glycimeris* (No. 101) with prominent umbones, almost bilaterally symmetrical valves with strong radial ribbing and internally a well marked *taxodont dentition* composed of a number of simple teeth, together with a great variety of forms, whose *heterodont dentition* consists of only a few, but tightly interlocking teeth. Such heterodont lamellibranchs include species of *Crassatella* (No. 84), *Venericor* (No. 108) with prominent radial ribbing and strong forwardly directed umbones, together with rather more delicate concentrically ribbed shells belonging to genera such as *Astarte* (No. 107) and the many genera that were once grouped together as *Venus* (No. 85). Burrowing forms include many species of *Phodadomya* and *Panopea* (No. 95), razor shells such as *Solen*, whilst pieces of driftwood bored by *Teredo* (No. 96) are quite commonly to be found in the marine clays. Sessile forms include *Ostrea* (No. 91) and *Lopha* (*Alectryonia*) (No. 93), but not the gryphaeas or exogyras of the Mesozoic. *Chama* (No. 106) was fixed by the prominent curved umbo of the larger valve, whilst varieties of mussels, such as *Modiolus* (No. 89) are other fixed forms.

The Tertiary rocks laid down under non-marine conditions yield species of *Cyrena*, *Corbicula* (No. 109) and *Ostrea*, but as at older horizons, the lamellibranch fauna of such beds shows little variety, consisting of thousands of individuals all belonging to one or at the most, two or three different species.

7e **Gastropoda**

Gastropods include snails, whelks, limpets, etc., as well as forms without hard parts, the slugs. They have a distinct head with a pair of eyes, often borne on stalks and a mouth with a *radula*, a rasp-like structure carrying rows of minute chitinous teeth. The majority of gastropods crawl sluggishly about on their sole-like foot. The visceral mass has been twisted so that it lies above the foot and points upwards and backwards. In the majority of the gastropods the visceral mass is protected by a univalve calcareous shell into which the head and the foot can be withdrawn. The aperture of the shell can be closed by a horny *operculum* (Fig. 25).

Most marine gastropods are benthonic, though some are pelagic and many of the benthonic forms have a pelagic larval stage. Gastropods can also live in freshwater and on land. Terrestrial gastropods are herbivorous, as gardeners realize when bewailing the damage done by slugs and snails. Many marine gastropods are likewise herbivorous, feeding on seaweed and lichens, but others are carnivorous. Certain of these such as *Buccinum* (the whelk) use their radula to bore through the shells of other molluscs and then protrude their long proboscis through the hole to extract the soft parts, others such as the marine snail, *Natica*, secrete chemicals to aid the boring process. Other gastropods such as *Vermetus* are sessile, feeding on microscopic plankton.

It is convenient here to recognize two main groups of gastropod shells, those which are coiled and those which are uncoiled. This division is an arbitrary one based on one category only, shell morphology, and cuts across the classification used by zoologists based on differences in the soft parts.

Gastropods with uncoiled shells include the benthonic limpet-like forms and also the pelagic pteropods. In the cap shells or *limpets* the shell is a low cone with a circular or elliptical opening. Limpets commonly live in very shallow water and clamp themselves so tightly to the rocks that they can withstand exposure between tides. When the tide flows the limpet will move off to feed and then return to its 'home', where its shell exactly fits on to the rock surface. Certain cap-shells, e.g. *Emarginula* (No. 116) have a narrow slit for the protrusion of the anus, whilst in the 'key hole' limpet *Fissurella*, there is an opening in the apex of the shell, out of which water can flow after passing over the gills and past the anus.

The *pteropods* are pelagic gastropods, swimming by means of the greatly expanded and very thin foot. Modern pteropods have a very thin calcareous shell in the form of a cone with almost parallel sides. They rarely exceed half an inch in length, but in the upper waters of some oceans occur in such abundance that the shells sink to the sea floor to form *pteropod ooze*. Hyolithids and tentaculids, common fossils at certain horizons in the Lower Palaeozoic rocks, have been claimed as pteropods. The delicate tapering shells (Fig. 12), without an opening at both ends as in the scaphopods (p. 140), certainly resemble recent pteropods, but there are no records of similarly shaped shells from the Mesozoic strata. The Palaeozoic conularids, now regarded as medusoids (p. 125), used formerly to be grouped with pteropods and it is indeed possible that the hyolithids and tentaculids are not related to the modern pteropods.

The coiled gastropods show a great deal of variation. Species of *Turritella* (No. 110) are common fossils at many horizons in the Eocene rocks. These shells are in the form of a very acute angled cone with the *spiral angle* between the opposite sides of the cone little more than 10°. At the apex of the cone, the first formed shell or *protoconch* may be preserved. Each turn of the spire is a *whorl*, and the line along which the whorls are in contact is the *suture*. The shell is ornamented by a number of delicate ridges parallel to the suture and therefore described as longitudinal. At the widest part of the shell is the *aperture* into which the creature can retract its head and foot. If the shell is held with the apex upwards, the aperture is on its right-hand side for the coiling is *dextral* or right-handed. It is the English convention to hold the apex pointing upwards, but in many French books the apex is pointing downwards. The aperture of the shell is oval, the lip is not notched and therefore it is described as *holostomatous*. If a section is cut down the axis of the spire, or if there are large enough breaks in the shell wall, it will be seen that the inner faces of the whorls are united to form a solid *columella*, so *Turritella* is described as *imperforate*.

Turritella has been chosen as a fairly simple type of gastropod. It can be described concisely as a holostomatous imperforate turreted form, thinking in

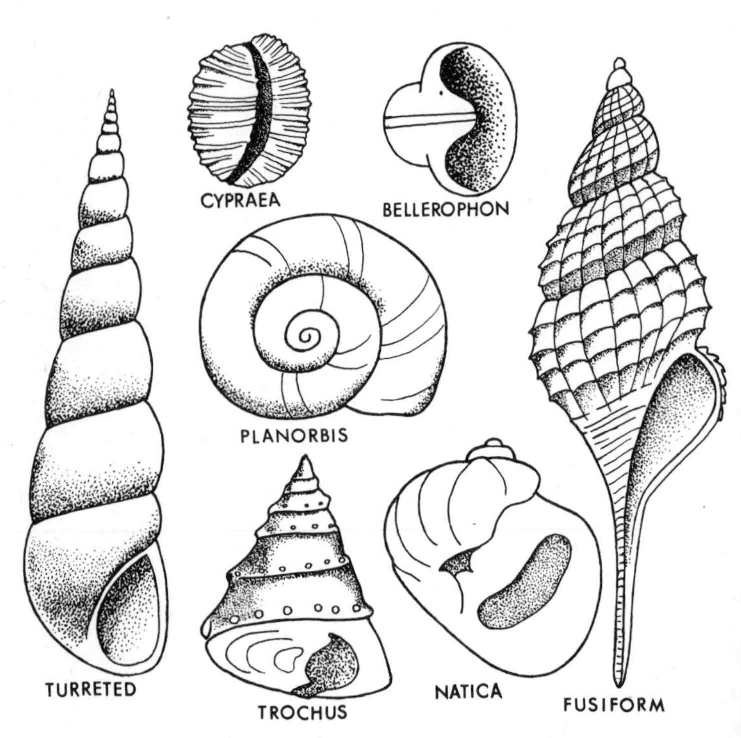

Fig. 26. *Variations in the coiling of gastropods.*

terms of three morphological features – form of aperture, presence or absence of columella and type of coiling. *Perforate* forms are those in which the whorls are coiled around a central cavity or umbilicus. In *siphonostomatous* gastropods, e.g. *Cerithium* (No. 110) the walls of the aperture are notched for the protrusion of either an anterior canal or for both anterior and posterior canals to carry the anterior inhalent and the posterior exhalent siphons. The coiling is right-handed in the majority of gastropods, much more rarely left-handed or *sinistral* as in *Neptunea contraria* (No. 113). Ornamentation varies from those with smooth shells to those with very strongly ornamented shells. The ornamentation can be delicate longitudinal lines as in *Turritella* or these can be much stronger, culminating in forms such as *Pleurotomaria* (No. 115) with a prominent slit band which has different ornamentation from the rest of the shell. There may be transverse features, parallel to the edge of the aperture, varying from delicate growth lines to strong ribs, tubercles or spines. Many of the more strongly ornamented forms, e.g. *Murex* (No. 120) show well-marked *varices* marking periods when there was a pause in the longitudinal growth of the shell, so that the apertural margin with its associated spines, ribs, etc., became thickened. Finally there is great variation in type of coiling. The chief types are listed below:

Capuliform as in the slipper limpet *Capulus* with the apex of the shell excentric and only the part of the shell near the apex showing slight coiling.

Convolute as in the cowrie, *Cypraea* (Fig. 26) with the outer whorl completely enveloping and hiding the spire.

Discoidal as in *Planorbis* (Fig. 26) with the whorls coiled almost in a plane spiral.

Fusiform as in *Fusus* (Fig. 26) with the anterior canal so well developed that the shell becomes spindle sloped.

Naticiform as in the sea snail *Natica* (Fig. 26) with a globose outer whorl and a small spire.

Planispiral as in *Bellerophon* (Fig. 26).

Trochiform as in *Trochus* (Fig. 26) with a flat base and a flat-sided cone with apical angle of 45° or more.

Turbinate as in *Turbo* with a convex or rounded base and conical sides.

Turreted with very small apical angle.

The Geological History of the Gastropods

In the Palaeozoic rocks gastropods are less common and less important than are the lamellibranchs. Forms found in the marine strata include the turreted *Murchisonia* of the Devonian and Carboniferous, the discoidal *Euomphalus* of the Carboniferous Limestone, the trochiform *Pleurotomaria* (No. 115) with its prominent slit band and the globular *Bellerophon* (Fig. 26) of the Carboniferous Limestone, which might be mistaken for a cephalopod, but for its well-marked slit band and, of course, absence of suture lines. The non-marine beds around the Silurian–Devonian boundary yield gastropods such as *Platyschisma* (No. 117), but unlike the lamellibranchs, gastropods are rare in the Coal Measures.

In the Mesozoic rocks gastropods are slightly more common and include the first siphonostamous forms. Turreted forms such as *Aptyxiella* (No. 3) are fairly common in some of the Jurassic limestones, together with species of *Pleurotomaria*. The 'Purbeck' or 'Sussex' Marbles of the Purbeck and Wealden Beds are shell beds of the pond snail *Viviparus* (Nos. 111–112), whilst gastropods are not too uncommon in the shallower water beds of the Chalk, such as the Lower Chalk and the Chalk Rock, but are much rarer in the purer and finer-grained chalks that were laid down in deeper water.

The Tertiary rocks yield gastropods in great variety and abundance, even more so than lamellibranchs. As already mentioned (p. 152) the Tertiary rocks are largely correlated on a basis of these lamellibranch-gastropod assemblages. Beds laid down on land may contain the snail *Helix*, in the freshwater beds *Lymnaea* (No. 122) with its thin spiral shell, very large last whorl and rounded aperture and the discoidal *Planorbis* (Fig. 26) may occur in sufficient numbers to form beds of soft limestone. A common fossil of beds of brackish water origin is the strongly ornamented turreted siphonostomatous gastropod *Potamides* (No. 123). Marine beds laid down in littoral waters yield *Emarginula* (No. 116), *Fissurella* and other limpets, together with *Trochus* (Fig. 26), the winkle *Littorina* (No. 124), etc. Beds deposited further off-shore contain a great variety of turreted forms, species of the holostomatous *Turritella* or of the siphonostomatous *Cerithium* (No. 110), the elongate *Fusus* (No. 121), together with strongly ornamented forms such as *Murex* (No. 120) and *Voluta* (No. 119) or species of *Aporrhais* (No. 118) with greatly expanded outer lips. The gastropods have clearly not yet reached their acme.

8 ECHINODERMATA

The echinoderms (Greek *spiny-skinned animals*) are highly organized and exclusively marine. Their skeletal parts are composed of calcite laid down in crystalline form and possessing a well-developed cleavage parallel to the faces of a rhombohedron. Broken fragments of echinoderms are therefore easily recognizable (Fig. 27). In addition, the skeletal parts have a distinctive honeycomb structure, clearly visible under the microscope.

Morphologically the echinoderms are so different from other phyla that comparisons can only be made in the larvae stage. It is then seen that the echinoderms are more closely related to the chaetognath worms, the hemichordates (p. 172) and the vertebrates than to the other phyla of the invertebrates.

A convenient subdivision of the phylum is as under:

	CLASS
Pelmatozoa (attached forms)	Cystoidea (extinct) Blastoidea (extinct) Crinoidea (sea lilies) Edrioasteroidea (extinct)
Eleutherozoa (free-moving forms)	Echinoidea (sea urchins) Stelleroidea (star fish) Holothuroidea (sea cucumbers)

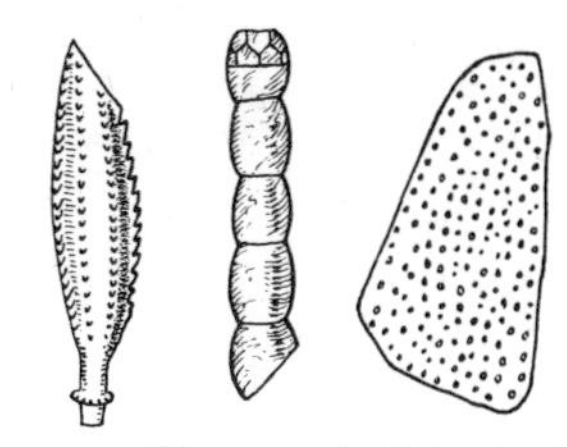

Fig. 27. *Cleavage and cellular structure in echinoderms. On left – Broken spine of a regular echinoid. In centre – Bourgueticrinus, a crinoid found in the Chalk, with a broken stem ossicle. On right – Magnified fragment of an echinoderm showing characteristic 'honeycomb' structure.*

Pelmatozoa

8*a* The **Cystoids** are a primitive group restricted to certain beds of the Lower Palaeozoic strata. Their body was sac-like, composed of many plates (Fig. 28), usually irregularly arranged, though in the later forms there was a tendency towards radial symmetry. Certain of the plates are penetrated by regularly arranged pores, which must have been used for respiration, whilst there was also a system of food gathering grooves connected to the mouth on the upper surface.

8*b* The **Blastoids** (Ordovician to Permian) were more highly organized. Their cup or *calyx* was made up by a limited number of plates showing a pronounced five-fold symmetry. The mouth and anus were on the upper surface. The plates of the upper part of the calyx were notched by five *ambulacral areas*, each of which bore a large number of food gathering brachioles, and were underlain by a complexly folded respiratory system, bathed by sea water which entered through pores between the brachioles (Fig. 28).

Normally blastoids are rare fossils, but at some horizons in the limestones of Carboniferous age these small rather bud-like echinoderms can be found in great numbers.

8*c* The **Crinoids** are much more common fossils than are either the cystoids or the blastoids. In a typical crinoid (Fig. 29) the *calyx* or cap is carried on a long somewhat flexible *stem* composed of many *ossicles*, circular or sometimes five-rayed in cross-section, and penetrated by a central opening for the passage of nervous organs. The stem terminated sometimes in an anchor-like structure, other

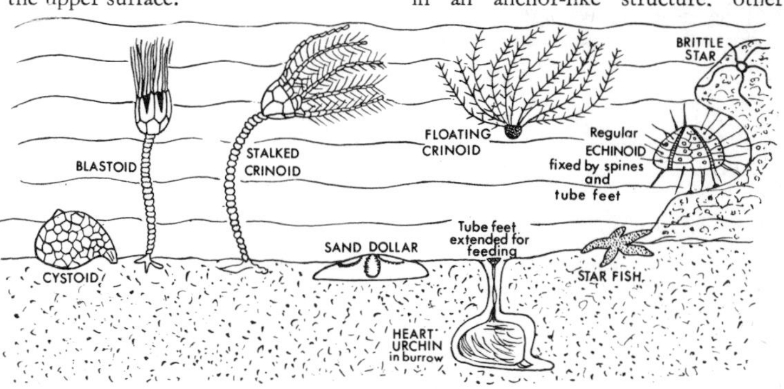

Fig. 28. *Selected echinoderms in their position of life.*

forms attached themselves to the sea floor by *cirri*, other crinoids cemented the base of the stem to the sea floor. To the upper surface of the cap were attached the five food gathering *arms*, which usually branched and sometimes bore *pinnules* (No. 126). Each arm and pinnule were made up of *ossicles* with a certain degree of movement on each other. Down each arm and pinnule ran a food groove lined by cilia whose movement produced a current carrying microscopic food particles to the central mouth on the upper surface of the calyx. In certain of the pinnulate crinoids the combined length of the food grooves was as much as a quarter of a mile. Each arm was attached at its base to one of the five *radial* plates which made up the upper ring of the calyx (Fig. 29). Below the radials, and alternating with them, was a

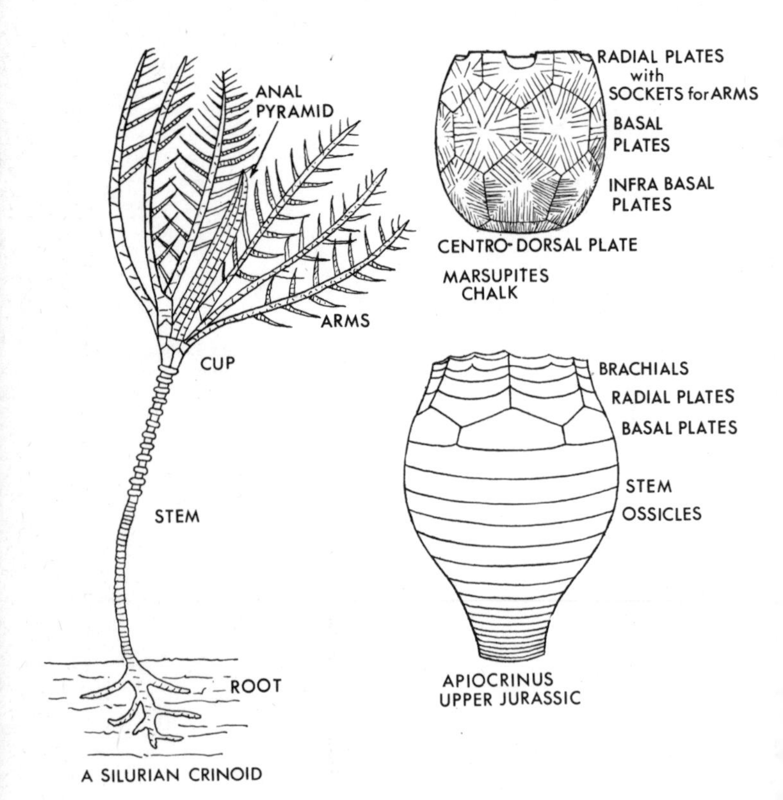

Fig. 29. *Selected crinoids.*

circle of five *basal* plates and below this, in some crinoids, another circle of five *infrabasal* plates, which were in sequence, but not in contact, with the radial plates. It is only the *dicyclic* crinoids, which have a recognizable ring of infrabasals in their cup. In the *monocyclic* crinoids (No. 129), only basals can be seen and even these are sometimes difficult to recognize. In certain crinoids, the five-fold symmetry of the plates of the cup is modified by the addition of an *anal* plate. The cup enclosed the digestive system. The upper (ventral) surface of the cup, the *tegmen*, is usually covered by a large number of small irregularly arranged plates. If five double rows of interlocking plates in sequence with the arms are recognizable, these are called the *ambulacral* plates, those between are *interambulacral* plates. In one group of Carboniferous and Permian crinoids, the *camerate crinoids*, the mouth and food grooves were covered by a convex rigid tegmen. As the basal parts of the arms are also incorporated in the calyx (No. 127), it is often a difficult matter to work out and name correctly the various plates making up the calyx of camerate crinoids. Use the same procedure as with any crinoid. Look for the plates where the arms articulated on to the cup, then the lowest plates in

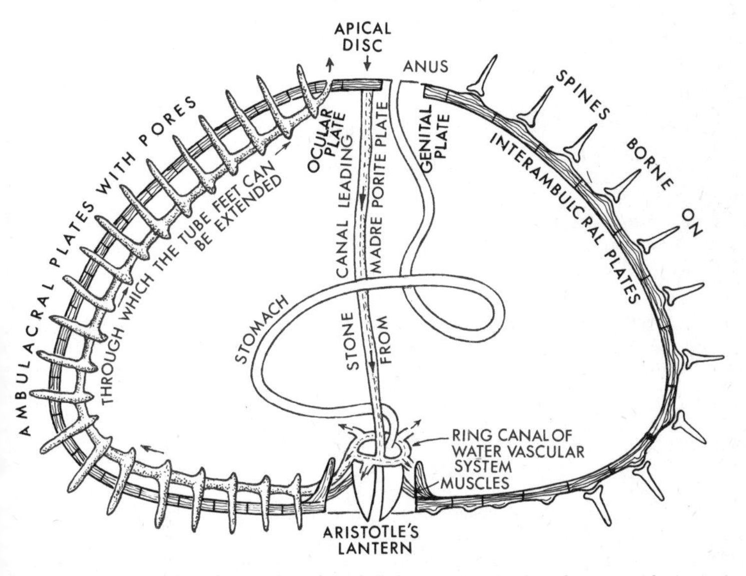

Fig. 30. *Section through a regular echinoid. Soft parts in red – Hard parts in black.*

sequence with the arms must be radials, with the alternating basals forming the ring below.

The Geological History of the Crinoids

The Crinoids range from the Ordovician to the Recent, but they were at their acme during the Upper Palaeozoic. Beds of limestone, often a score or more feet in thickness, are made up of fragments of the arms and stalks of crinoids (No. 125). Complete cups with part, at least, of the arms and stalk attached are much more uncommon. During the Upper Palaeozoic, particularly during the Carboniferous Period, the crinoids showed great morphological diversity. As well as the camerate crinoids mentioned above, two other sub-classes can be recognized, the *inadunate crinoids* with the plates of the calyx firmly joined together and the arms free above the radials and the *flexible crinoids* with the plates of the calyx not all rigidly joined whilst the mouth and food grooves were not covered. The great 'forests' of crinoids, whose remains produced the beds of crinoidal limestone, must have lived in fairly shallow water. All these sub-classes died out at the end of the Palaeozoic Era. The post-Palaeozoic crinoids belong to the Articulata with a small simple cup and pinnulate arms. Certain of the *articulate crinoids* abandoned their stem after the early larval stage and became free swimming with a centrodorsal plate supporting the infrabasals. In some Cretaceous forms the thin arms were up to 4 feet in length. Certain of these pelagic crinoids, such as *Marsupites testudinarius* (Fig. 29) are valuable zone fossils. In the modern seas crinoids are represented by pelagic forms such as *Antedon* and a variety of benthonic forms ranging from very shallow water to the ocean depths, but they are a very insignificant element of the marine fauna compared with their acme in Upper Palaeozoic times.

The **Edrioasteroidea** are a very rare group of Palaeozoic echinoderms with a sac-like body made up of many irregularly arranged polygonal plates. From the central mouth, extended five curved and unbranched ambulacral areas, which did not bear brachioles.

Eleutherozoa

8d The **Echinoids** (sea urchins) are entirely marine. Their skeletal structures are amongst the most complex of those found in the invertebrates. The sea urchins are divided into two sub-classes, the *Regularia* dominated by a five fold (pentameral radial symmetry) and the *Irregularia* with bilateral symmetry superimposed to varying degrees on the basic radial symmetry.

In Mesozoic to Recent regular echinoids the test is composed of twenty rows of calcareous plates (Nos. 130–131). There are five double rows of large interlocking plates, the *interambulacral areas*, alternating with five double rows of much smaller plates, the *ambulacral areas*. On the interambulacral plates are the sockets of stout *spines* (Fig. 30), which were longer in many regular echinoids than the diameter of the test. Smaller spines were attached to both the ambulacral and the interambulacral plates. Along the outer margin of the ambulacral plates occur a row of *pores*. These are part of the *water vascular* or water-circulating *system*, which is a distinctive feature of all echinoderms, but is

plates of the test were thin and as they overlapped each other like tiles on a roof, there was a certain amount of flexibility in the test, which was also not protected externally by long spines. These Palaeozoic forms must have lived in fairly quiet waters for they certainly could not have survived on the rocky wave-battered coasts inhabited by many existing regular echinoids. The cidaroids are the one Palaeozoic group which have survived to the present. They have a strong test composed of 20 rows of plates in all and are protected by strong stout spines, borne singly on each ambulacral area (No. 136). From a cidaroid stock have developed the considerable variety of regular echinoids found in the Mesozoic and Tertiary rocks.

In the *Irregular Echinoids* the periproct is not within the apical disc, but has moved into the posterior interambulacral area. In the less advanced forms (e.g. *Clypeus* (No. 134), the periproct is only just outside the apical disc, but in the more advanced forms (e.g. *Micraster* (No. 132) it has moved completely off aboral surface and is on the margin of the test, whilst in other forms it is on the oral surface. The peristome may also be excentric. In forms such as *Clypeus*, the development of bilateral symmetry is limited to the position of the periproct, whilst the peristome is slightly eccentric. In the more advanced *spatangoids* (heart-urchins), e.g. *Micraster*, bilateral symmetry is much more pronounced, the anterior ambulacral area is markedly different from the other four, whilst in some forms the apical disc is elongated. A further change is in side view. It is exceptional for regular echinoids to show much flattening on the aboral surface, but this is much commoner amongst the irregular echinoids and reaches its acme in the *clypeastroids* or sand-dollars. At the same time the ambulacral areas become markedly petalloid with the pore-bearing portion restricted to the upper part of the aboral surface (No. 135). Another feature is the reduction in the size of the spines. In the heart-urchins and the sand-dollars, they are reduced to a hair-like covering.

Early in Mesozoic times the Irregularia developed from a cidaroid-like ancestor and began to inhabit a much greater variety of environments (Fig. 28), particularly those with soft bottoms. The heart-urchins use their spade-like spines to burrow into sands. The burrows are connected to the sea floor by a mucus-lined canal. Fresh sea-water is drawn in and stale water expelled through the canal. The long tube feet of the anterior ambulacral area also extend up the canal. The heart-urchins are 'deposit feeders', their tube feet picking over the sea floor around the mouth of the burrow and drawing down to the mouth any edible fragments. The sand-dollars lie just buried beneath the sediment. Any fragments of organic matter are carried along cilia-lined furrows on the oral surface to the mouth. As a result of feeding on such small food particles, both the heart-urchins and the sand-dollars have lost the complex Aristotle's Lantern of the scavenging regular echinoids and the less advanced irregular echinoids. The test of the heart-urchins is very thin and easily crushed, but that of the sand-dollars is often strengthened internally by vertical partitions, for many of these urchins inhabit the intertidal zone and have to be able to withstand the downward blow of breaking waves.

Both regular and irregular echinoids are common in limestones of Mesozoic and Tertiary age (Nos. 132–137). At some horizons, notably in the Cretaceous Chalks, echinoids are used as zone fossils. They are conspicuous and easily recognizable. Furthermore, even broken fragments can be assigned to a definite horizon as, for example, incomplete specimens of *Micraster* in the Upper Chalk of England and France or the spines of *Cidaris* in the Danian Chalk of Denmark. Echinoids are rare in the marine clays that predominate in the Tertiary deposits of northern Europe, but are abundant, particularly the clypeastroids, in the shallow water limestones and sands of southern Europe.

8e Although the **Stelleroids** (starfish) range from the early Ordovician to the Recent, it is extremely unusual to find complete specimens as fossils. The skeleton was not a rigid box as in the sea urchins and the sea lilies, but was composed of many calcareous ossicles, which fell apart after the decay of the soft parts.

Starfish beds occur at a number of horizons, but the term does not mean that these particular beds are replete with starfish remains, rather that specimens (No. 138) may be found in them often only after prolonged search.

From the central disc extend five arms, long and flexible in the *brittle stars*, strong and stout in the *sea stars* (Fig. 28). The water vascular system (p. 160) is well developed with a madreporite on the upper surface and powerful sucker-like tube feet along the ambulacral grooves on the underside of each arm. The sea stars crawl slowly over the sea floor or climb rocks by means of their tube feet. They are exceedingly voracious, preying particularly on lamellibranchs. They wrap their arms round the shell of an oyster and then exert the pull of their tube feet against the closing muscles of the shellfish. Eventually the oyster will be forced to open its shell. Then the starfish extrudes its stomach through the opening and digests the soft parts of the oyster. Oyster beds may be ruined if they are attacked by starfish.

8f **Holothurians** (sea cucumbers) are leathery creatures, whose skin contains many small calcareous spicules of diverse shape – anchors, wheels, crosses, etc. (Fig. 15). Like the skeletal parts of other echinoderms, each spicule is optically a crystal of calcite. This property is useful in distinguishing holothurian spicules from other microfossils. Impressions of holothurians have been found in the Middle Cambrian Burgess Shale of British Columbia and also in Upper Jurassic Lithographic Stone of Solenhofen in Bavaria.

9 ARTHROPODS

Approximately three-quarters of all known vertebrates and invertebrates are arthropods, the most highly organized phylum of the invertebrates. The distinctive features of the arthropods (*joint-footed*) are first the bilateral symmetry of the segmented body and secondly that each body segment and also the segments which may be fused together in the head or tail region, bear a pair of jointed appendages. The appendages normally function as walking or swimming legs, but some of them may be modified to form wings, antennae, jaw-like mandibles or for other special functions. The soft parts, body and appendages are protected by a chitinous covering secreted by the ectoderm or outer skin of the animal. This hardened covering or *exoskeleton* not only protects and supports the soft tissues, but also provides attachment for the muscles. Arthropods develop from an egg and then pass through a larval stage or stages before attaining the adult form. The familiar sequence caterpillar–chrysalis–butterfly shows how different the stages of an individual arthropod's life history or *ontogeny* may be. Unlike our own internal skeleton of bones the arthropod's exoskeleton cannot grow steadily and gradually. Once formed the exoskeleton is complete and rigid, and therefore periodically, as the soft parts increase in size the exoskeleton must be shed. It splits along *sutures* or special lines of weakness, is discarded and then a new and larger exoskeleton is quickly formed. In the case of certain fossil arthropods it has been possible to reconstruct the successive growth stages by finding a sequence of young individuals or of their moults.

The Arthropods can be divided into the following main groups:

(*a*) Insecta.
(*b*) Chelicerata (spiders, scorpions, mites, etc.).
(*c*) Crustacea (lobsters, crabs, ostracods and barnacles).

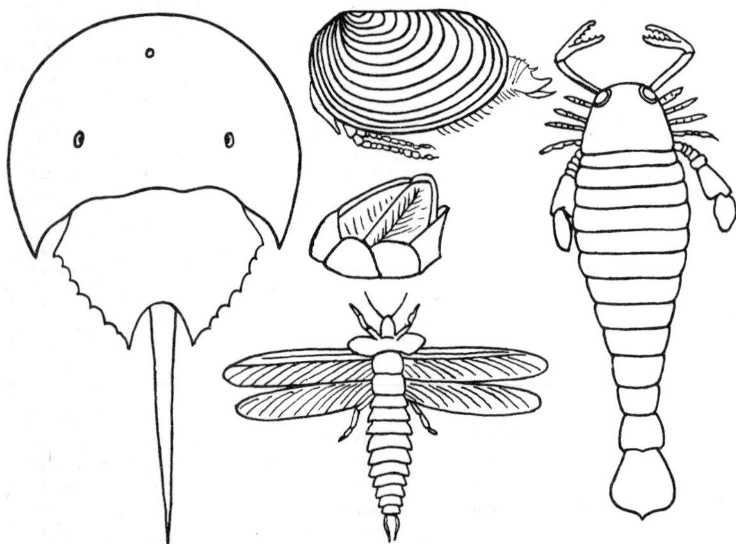

Fig. 31. *Representative arthropods. Left – Limulus, the King Crab. Top centre – A branchiopod. Middle centre – Cirripede, a barnacle. Bottom centre – An insect from the Coal Measures. Right – Pterygotus, an eurypterid.*

(*d*) Myriapoda (centipedes and millipedes).
(*e*) Trilobitomorpha.

The extinct Trilobitomorpha, including the trilobites, are a well-known and important group of fossils. The majority of the other groups are either not found fossil or are only preserved under rather exceptional circumstances.

9*a* Insecta

The **Insects** (*cut into*) take their name from the clear-cut divisions, which separate their body into head, thorax and abdomen. Modern insects are extremely numerous and varied, but many of them are minute in size, also their bodies are often delicate. For these reasons it is not surprising that the fossil remains of insects are only known from a limited number of horizons and also that special techniques may be needed for their extraction. The term 'Insect Bed' or 'Insect Limestone', which one sometimes comes across in the literature, usually means a bed in which insect remains have been found, rather than one in which they are abundant and easy to collect.

The oldest known insects occur in beds of chert of Middle Old Red Sandstone age at Rhynie in Aberdeenshire, Scotland. The same beds are famous for containing very early land plants (p. 194). The chert originated as a peat bog which was permeated by water rich in silica derived from nearby hot springs, so that both the swamp flora and the insects living in the swamp have been preserved. They are wingless forms, some related to the modern springtails.

The Coal Measures have yielded numerous winged insects (Fig. 31), including ancestral dragonflies with a wing span of over 2 feet. Their wings were permanently open and could not be folded back over the abdomen as in modern dragonflies. Other insects from these beds include cockroaches and cricket-like forms.

In the Mesozoic and Tertiary rocks the best-known horizons yielding insects are certain 'Insect Beds' in the British Lias, the Oligocene beds of the Baltic coast containing the well-known spiders preserved in amber (No. 139) and a Miocene lake-deposit at Florissant in Colorado. Fine-grained ash from volcanic eruptions fell into the lake, carrying down and entombing the creatures that flew over and swam in the lake.

Clearly from what has been said above, insect remains are far too infrequent to be used for chronological purposes, but within the last few years it has been shown that fossil insects can give valuable help in the study of the deposits laid down during the oscillating climatic conditions of the later part of the Pleistocene Period. Fragments of beetles and other insects can be isolated from peat and lake silts and examined under a binocular microscope. Specimens identical with those of modern species have been obtained from beds whose age is estimated at about half a million years. If we assume, and the available evidence justifies the assumption, that the species found in fossil faunas had the same climatic and ecological requirements as their modern representatives, then in specialist hands the study of fossil entomology is proving a valuable tool in deducing within narrow limits the climatic conditions under which certain Pleistocene deposits were laid down.

9b Chelicerata

The distinctive features of the chelicerates are that one pair of the anterior appendages are pincers (*chelea*), whilst the pre-oral segments are fused with some of the post-oral segments to form a rigid *prosoma* or cephalothorax.

The Chelicerata can be divided into the arachnids, the terrestrial spiders and scorpions, the merostomes, aquatic forms including the modern King Crab, *Limulus*, and the extinct eurypterids (Fig. 31).

The earliest scorpions recorded are of Silurian age, whilst spiders range back to the Devonian. They are, however, extremely uncommon. One interesting exception to the normally poor chances of preservation of terrestrial forms is provided by the remains of scorpions found in rocks of Triassic age near Bromsgrove in Worcestershire. The scorpion fragments occurred in sandy shales. These were dried, then plunged into hot water so that as they crumbled the chitinous fragments could be removed with a needle or a brush. The fragments were so dismembered that it was suggested that these scorpions had been broken up by other scorpions. If true, a case of cannibalism fossilized.

Merostomes have been found, at intervals, from the Cambrian upwards. The *xiphosurans* (sword tails) are represented today by *Limulus*, the King Crab. The King Crab has a large semicircular prosoma articulating with the opisthosoma and this with the long telson or tail spine. Modern King Crabs live in shallow waters along the east coast of North America, in the western Pacific and the Indian Ocean. They crawl ashore to breed. Slabs of the famous Lithographic Stone from Solenhofen in Bavaria have been found with specimens of *Limulus* preserved at the end of a trail of footprints. Presumably these had crawled too far from the lagoon to be able to return to it before being desiccated by the heat of the sun.

The *Eurypterids* are only found in Palaeozoic strata. They were the largest of the arthropods, for specimens of *Pterygotus* are known up to 10 feet in length. Beneath the prosoma were six pairs of appendages, the anterior ones effective pincers, then four pairs of short walking legs, whilst the last pair was larger and oar-like. There were about twelve segments in the abdomen, whilst the metasoma terminated in a spike-like telson (Fig. 31). The streamlined shape of the body and the oar-like pair of appendages suggest that the eurypterids were good swimmers. The majority of specimens come from beds around the Silurian–Devonian boundary in the British Isles, Scandinavia and the United States. They are not found in beds yielding such normal marine forms as trilobites and corals, but only in strata, which from other evidence were probably laid down in water of unusually high or unusually low salinity, that is in semi-enclosed lagoons. As completely articulated carcases have been found, it is clear that the eurypterids must have lived in this environment and were not swept along after their death down the rivers flowing into the lagoons. Agnathids (p. 177) are often found in the same beds as eurypterids, so it is possible that the eurypterid's effective looking pincers were used for preying on these heavily armoured fish-like creatures.

9c Myriapoda

Though they range well back into the Palaeozoic, the *myriapods* are extremely rare as fossils and therefore will not be considered further.

9d Crustacea

The *Crustaceans* are mainly aquatic arthropods with biramous (two branched) appendages rising from a basal joint. The inner appendage, the *endopodite*, is jointed and used for walking, whilst the outer, the *exopodite*, functions for respiration and swimming. The chief classes of crustaceans found fossil are the ostracods, the branchiopods, the sessile cirripedes, barnacles and certain malacostracous crustaceans including the crabs and lobsters.

The body and appendages of the *Ostracods* are enclosed in a bivalved calcareous shell. They are usually minute, up to a few millimetres in length, though some 'giants' may range up to an inch. Aquatic forms, they occur in marine, brackish or freshwater deposits. At some horizons, such as in the Wealden Beds, they occur in myriads. There is very considerable variation in the shape and particularly in the ornamentation of the valves (Fig. 15). Genera and species have a limited vertical range, hence the ostracods are of value as zonal fossils. Owing to their minute size, they usually have to be examined under a binocular microscope. At many horizons they are a valuable part of the microfauna obtained from shales, marls and limestones penetrated by oil wells.

The *Branchiopods*, including the water fleas, also have a bivalved carapace, rather resembling a small lamellibranch shell, though there is no hinge structure and the shell is made of chitin and not of calcium carbonate. At certain horizons, notably in the Triassic and Rhaetic beds, bedding planes may be covered with the carapaces of *Euestheria* (No. 141), but apart from this, the branchiopods are of little geological importance.

The *Cirripedes* (barnacles) have a free swimming larval stage, but finally the larva settles on some hard object. It attaches itself by the head region, grows a number of thin calcareous plates for protection, whilst the feet become modified into delicate curved appendages used for conveying food to the mouth, hence the name cirripede (*curl foot*). Fossil cirripedes range back to the Palaeozoic, but were never an important group. One may sometimes find, as in the 'crags' of East Anglia, barnacles preserved on the shells of large molluscs (No. 113) or one may find isolated cirripede plates, recognizable by their triangular shape and delicate ornamentation (Fig. 31).

The *Malacostracous crustaceans* range back into the Cambrian, but are not a geologically important group. At certain horizons, such as the Gault and London Clay of England, one may hope to find easily recognizable fossil crabs (No. 140), the Lobster Beds of the Lower Greensand at Atherfield in the Isle of Wight are not misnamed, for specimens of *Meyeria* (No. 142) are not uncommon, the Solenhofen Limestone of Bavaria has yielded many beautifully preserved crabs and lobsters and at some horizons in the Purbeck Beds of highest Jurassic to Cretaceous age the isopod *Archaeoniscus* is extremely abundant.

9e Trilobitomorpha

The *Trilobitomorpha* includes that well-known extinct group the trilobites together with a number of trilobite-like forms, mainly known from the Middle Cambrian Burgess Shale of the Rocky Mountains, where they occur as flattened films on shale. We are only concerned here with the **Trilobites**.

A representative trilobite is *Calymene blumenbachi*, the 'Dudley Locust' or 'Dudley Bug', as it was named by the quarrymen who frequently found specimens of it when working the Wenlock Limestone of Middle Silurian age in the abandoned quarries now the site of the Dudley Zoo near Birmingham. *Calymene* is reasonably common in working and disused quarries at the Wren's Nest near Dudley, along Wenlock Edge in Shropshire and other exposures of the Wenlock Limestone in the Welsh Borderlands.

The hard parts preserved are of the back or *dorsal* surface, which carried an exoskeleton of chitin impregnated with calcium carbonate and calcium phosphate. Two well-marked longitudinal grooves separate a central or *axial* strongly arched portion from two flatter lateral or *pleural* regions. Hence the name Trilobite. There is also a transverse division into the headshield or *cephalon*, the segmented *thoracic* portion and the tailpiece or *pygidium* (No. 143).

The cephalon shows an axial raised portion, the *glabella*, with furrows on its margins. The glabella does not reach to the anterior margin of the headshield. On either side of the glabella are the cheeks with on each a slightly raised eye. On each cheek there is a thin suture (*facial suture*), running from the *genal angle* at the outer posterior margin of the cephalon, round the inner margin of the eye and

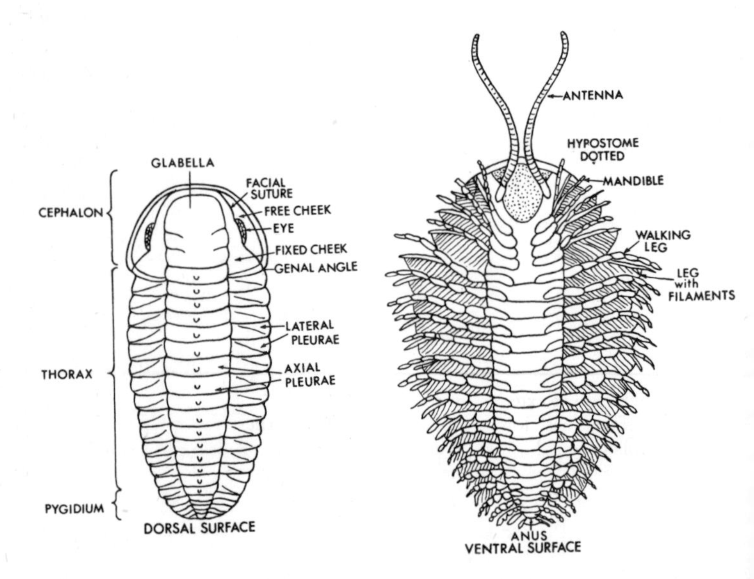

Fig. 32. *Dorsal and ventral surfaces of the Trilobite, Triarthrus.*

then forwards to the anterior margin. The facial suture divides the cheeks into the outer or *free* cheek and the inner or *fixed* cheek. A well-marked cross-furrow, the *neck-furrow*, extends from one fixed cheek to the other across the posterior part of the glabella. There are thirteen *thoracic segments*, all similar except that they become slightly smaller posteriorly. The pygidium consists of six segments fused together with the lateral parts showing increased backward curvature.

Some specimens of *Calymene* (No. 144) show the creature rolled up, like a wood louse, with the *ventral* (under) surfaces of the cephalon and pygidium in contact, the thoracic segments being able to articulate on each other so as to provide the necessary amount of movement.

Details of the undersurface of *Calymene* are not too well known, but they can be inferred from exceptionally well-preserved specimens of certain other trilobites, such as the partly pyritized specimens in very fine-grained mudstones of Lower Devonian age from the Rhineland, Germany, and of Ordovician age from New York State. Such specimens show that beneath the cephalon there were an anterior pair of antennae, behind these four pairs of two-branched appendages, which may have been used as mandibles for breaking up food and pushing it into the mouth. The hardened exoskeleton, which just extends on to the ventral surface, has joined to it a hardened plate or *hypostome*, lying in front of the mouth. Hypostomes may be found either in position or they may be detached. Beneath each segment of the thorax and the pygidium are a pair of jointed biramous appendages, consisting of an inner pair of walking legs and an outer pair bearing a fringe of filaments. The outer legs are interpreted as having functioned both for swimming and for respiration, the filaments supporting the gills. The anus is situated beneath the terminal segment of the pygidium (Fig. 32).

We can therefore interpret a 'typical' trilobite, such as *Calymene*, as a highly developed arthropod, capable of either walking over the sea floor or of swimming above it. It was probably a scavenger. Like other arthropods periodically it moulted, casting off its exoskeleton and growing a new one. The facial and other sutures were lines of weakness which facilitated the moulting. One often finds isolated free cheeks, which are clearly the result of moulting. The growth stages of *Calymene* are not well known, but for certain trilobites there have been found a complete ontogenetic series starting with a minute unjointed carapace representing the beginning of the cephalon. Then the thoracic segments developed, followed by the beginnings of the pygidium and so to the complete adult form. Such ontogenetic series have been obtained from the Shineton Shales of Shropshire, the Middle Cambrian rocks of Bohemia and Sweden as well as from horizons in the United States.

The Geological History of the Trilobites

The earliest known trilobites occur in rocks of Lower Cambrian age. These are highly developed forms, but we have no trace of their ancestors. They must have been soft bodied without any parts capable of fossilization. The Lower Cambrian trilobites are much more specialized than are the

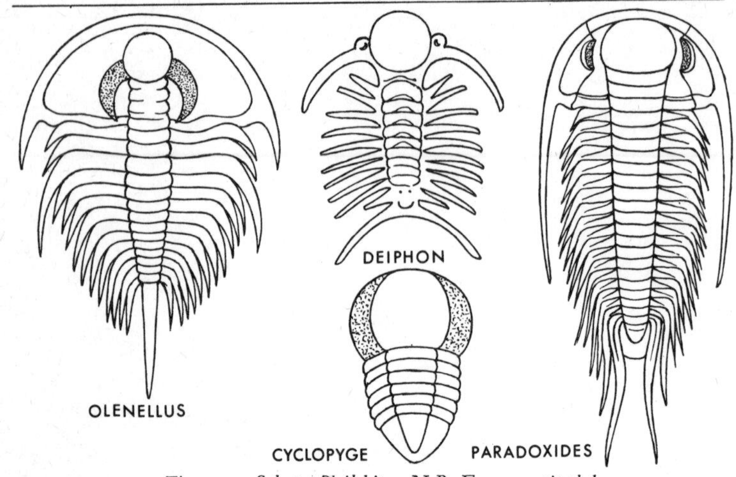

Fig. 33. *Selected Trilobites. N.B. Eyes are stippled.*

primitive chitinous inarticulate brachiopods found in the same rocks. It may well be that the waters of the Pre-Cambrian oceans were deficient in the mineral salts required by organisms for the secretion of hard parts. It is not until well up into the Cambrian succession that we find the fossil remains of animals with a calcareous skeleton, such as the archaeocyathids, early nautiloids, articulate brachiopods, etc. This may be a response to slight changes in the chemical composition of the sea water.

The Lower Cambrian trilobites are very distinctive. They include the *mesonacids*, such as *Olenellus* (Fig. 33), with a strongly spinose elongated body and long genal spines on the cephalon. They have either a very small pygidium or a long tail spine. There is no facial suture whilst the eyes are large and crescent-shaped. It is interesting that even at this remote time there is evidence of faunal provinces, for the Lower and Middle Cambrian rocks of Australia and Asia contain different assemblages from the olenellids and other forms found in Europe, North Africa and North America. The later Lower and Middle Cambrian beds yield other highly distinctive trilobites. Large *Paradoxides*, up to nearly 18 inches in length, as spinose as the mesonacids, but with a better developed pygidium, whilst the glabella expands anteriorly and there is a well marked facial suture (Fig. 33). The suture is *opisthoparian*, that is it runs from well behind the genal angle. In the same beds are to be found the minute *agnostids* and *eodiscids*. Blind forms without eyes or facial sutures, with only two or three thoracic segments and with the pygidium the same size as the cephalon (No. 145). The Middle Cambrian rocks of Britain and Scandinavia have been zoned in considerable detail on a basis of the limited vertical range of species of *Paradoxides* and *Agnostus*. It must be emphasized that prolonged search may be necessary to find even fragments of trilobites in the British Cambrian rocks, for at many localities the beds are so disturbed. In

Scandinavia, particularly in the Oslo neighbourhood, and in Bohemia, beds of the same age are much richer in well preserved trilobites (Nos. 11 and 145).

Many new families appear in the Upper Cambrian, especially in the Tremadocian, and range on into the Ordovician. These include the opisthoparian *Ogyiocarella* (Fig. 4) and *Basilicus* (Fig. 4) and the *proparian* calymenids with the facial suture in front of the genal angle. These are all 'normal' looking trilobites. In addition in the Ordovician, there are many *trinucleids* (Fig. 4) and (No. 146); blind forms with a large shovel-shaped cephalon terminating in long genal spines, the glabella is separated from the anterior margin by an ornamented border, there are few thoracic segments and the pygidium is small. Such forms are clearly adapted to feeding on nutritive matter whilst crawling through the muds of the sea floor. As well as these mud grubbers, there are a number of trilobites, such as *Ampyx*, *Staurocephalus*, and in the Silurian *Deiphon* (Fig. 33) with unusually thin exoskeletons, the eyes set near the margin, strongly inflated glabellas and also strongly spinose, all characters which would assist buoyancy. They are interpreted as nektonic forms. Additional evidence for a swimming form of life is the fact that such forms often occur in graptolitic shales, whilst the more normal trilobites are found in sands and silts which were clearly laid down closer to the land areas (Fig. 5). Trilobites are of great value in subdividing the beds of the sandy facies. The largest known trilobite, a lichid, over 2 feet in length, has been found in the Ordovician rocks of Portugal.

The trilobites reached their acme in the Ordovician and then began a slow decline. They are still important in the shallower water beds of the Silurian. In addition to the 'Dudley Bug', well known Silurian forms are strongly punctate encrinurids, such as *Encrinurus punctatus* with eyes on short stalks (No. 147), dalmanellids, such as *Dalmanites vulgaris*, another characteristic fossil of the Wenlock Limestone, with hundreds of lenses in its prominent eyes (No. 148). In the Ordovician nektonic form, *Cyclopyge* (*Aeglina*) (Fig. 33) the eyes are even more prominent, covering the whole of the dorsal surface of the cephalon except for the glabella.

Trilobites are still sufficiently numerous in the marine Devonian rocks, especially of the Rhineland, to be used for correlation purposes, but they were definitely on the decline. *Bronteus* (No. 149) with its distinctively shaped

Fig. 34. *Thecal variation in graptolites (after Bulman). (a) Dendroid graptolite to show mode of budding. Stolothecae uncoloured, stolon in grey, autothecae black, bithecae stippled. (b) Glyptograptus, Upper Ordovician. (c) Lasiograptus, Upper Ordovician. (d) Monograptus priodon with hooked thecae, Lower Silurian. (e) Monograptus convolutus, Lower Silurian. (f) Rastrites with strongly isolate thecae, Lower Silurian.*

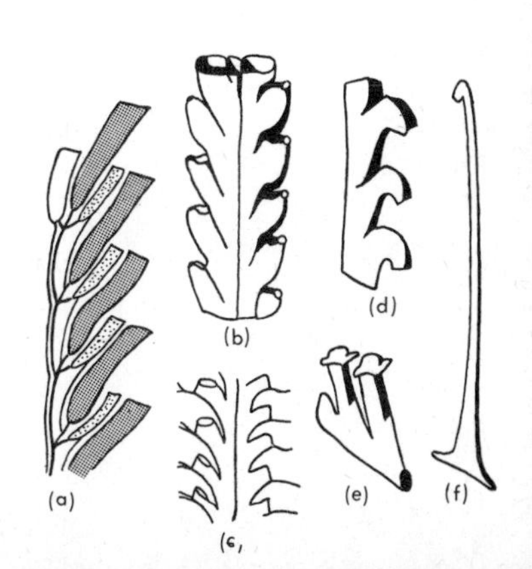

pygidium is a well-known form. Punctate specimens of *Griffithides* and *Phillipsia* are rather rare fossils in the marine Lower Carboniferous rocks of Europe. The last known trilobites are found in rocks of Permian age in the Isle of Timor and in the Salt Range of the Himalayas.

10 GRAPTOLITES

The **Graptolites** are one of the most important groups of fossils to be found in shales of Lower Palaeozoic age. They occur normally either as white films (No. 9) or as pyritized rods (No. 7). These have some resemblance to ancient writing, hence their name from the Greek *grapto*, write, *lithos*, stone. Most graptolites consist of one or more branches or *stipes* along which are a number of cup-like *thecae*. Because of this they were originally regarded as colonial organisms related to the hydrozoans. But this conclusion was based on specimens which had been flattened on the bedding planes. Later the technique was developed of dissolving in suitable acids pieces of chert or of limestone containing graptolites. The chitinous skeleton of the graptolites was not attacked, so uncrushed specimens could be floated off, mounted and sectioned. It was then discovered that the graptolites were not related to the hydrozoans or the bryozoans, but to the *pterobranchs*, a rare group of modern marine animals, including both benthonic and pelagic forms. The pterobranchs have gill slits and a notochord, a longitudinal skeletal rod, and therefore belong to the phylum of the Chordata, the same phylum as the vertebrates, though the pterobranchs are sufficiently distinct to be placed in a separate subphylum, the *Hemichordata*. Sections cut through certain uncrushed graptolites showed that the thecae are arranged in groups of three (Fig. 34). Each triad consisted of a large autotheca which probably housed a food-gathering zooid, a much smaller open bitheca whose function is doubtful, and lying along the line of the stipe a small stolotheca, which supported the base of the autotheca of the next triad. Through the line of stolotheca extended a stout chitinous tube, the *stolon*, comparable to that of the pterobranchs.

The earliest graptolites, found in beds of latest Cambrian age, are many-branched dendroid forms such as *Dictyonema* (Fig. 35). Its *rhabdosome* or colony is composed of many branches or *stipes* connected by short bars or *dissepiments*, which enabled the stipes to hang down in a bell-like form from the *sicula*, the initial thecal cup of the colony. From the sicula extended upwards a stout rod, the *nema*, with a flattened chitinous disk at its upper end. *Dictyonema flabelliforme* is of zonal importance, for it is widely distributed throughout a limited thickness of beds of the Tremadocian Stage in North America and Europe. It must have been planktonic and was believed to have attached itself by its disk to masses of floating seaweed. This view has recently been challenged and it is suggested that graptolites had their own buoyancy mechanism, perhaps gas bubbles or air sacs in the soft parts, which are quite unknown for only the chitinous skeleton has been preserved. Many branched *dendroid graptolites* persist with relatively little change from latest Cambrian to early Carboniferous times, when they became extinct. A number of the later dendroids, with a short stout nema, were clearly benthonic and not planktonic forms.

At the beginning of the Ordovician Period, the *true graptolites* developed from one branch of the Dendroidea. The bithecae and the chitinized stolon were lost, but a common canal running along the length of the stipe remained. Unlike the stable dendroids, the true graptolites showed rapid evolutionary change and hence great morphological diversity. The first change was a reduction in the number of stipes leading to *Tetragraptus* and finally the two-branched *Didymograptus* of the Arenigian Stage (Fig. 4). Associated with this were changes in the relation of the stipes to the nema from pendant forms such as *Dictyonema* and *Didymograptus bifidus* (Fig. 4) to horizontal forms such as *Didymograptus extensus* (Fig. 4), reclined forms such as *Dicellograptus*, and scandent forms such as *Glyptograptus* (Fig. 34). The scandent forms are *biserial* with thecae on both sides of the stipe in contrast to the *uniserial* arrangement in the dendroids and other true graptolites. In *Dicranograptus* (Fig. 4) the stipes near the sicula are biserial, the later formed parts uniserial. There are also changes in the shape of the thecae. In the dendroids and the lower Ordovician graptolites these are cylindrical, inclined at an angle to the stipe, but in the upper Ordovician graptolites there is much variation in thecal shape. This trend reaches its acme in the uniserial scandent *monograptids*, which are the dominant family of the Silurian rocks (Fig. 34). At the same time the number of thecae in a complete rhabdosome

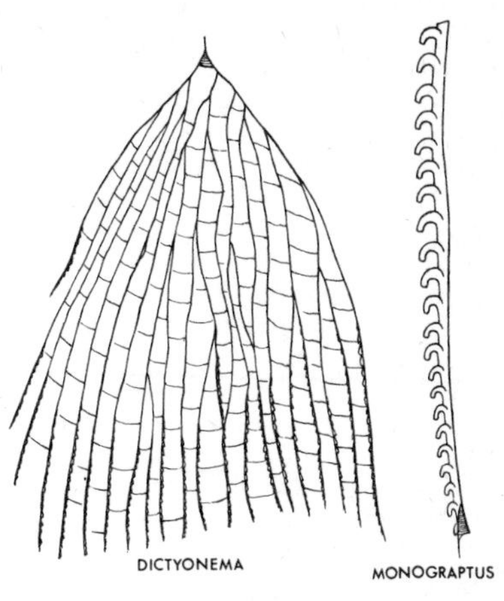

Fig. 35. *Contrast between colony of Dictyonema and a Monograptid.*

were reduced from the 20,000–30,000 in a large *Dictyonema flabelliforme* to 100–200 in a dicellograptid and finally to 10–20 in one of the last monograptids (Fig. 35). The many different species of monograptids are identified mainly on thecal form, but this will only be clearly apparent in specimens which came to rest with the thecae lying flat on the bedding plane (No. 7). If the stipes became twisted so that the thecae were pointing upwards then the thecae were pushed into the stipe.

In Britain the last of the true graptolites, the monograptids, became extinct in the latest Silurian rocks, in Bohemia they range slightly higher into strata of earliest Devonian age. The true graptolites are invaluable zonal forms in the shale facies of the Ordovician and Silurian Systems. Genera and species have a narrow time-range but a wide geographical distribution. Numerous forms, such as the 'tuning forks' of the Llanvirnian Stage (No. 9) and the strongly isolate *Rastrites* (Fig. 34) of the Upper Llandoverian are very distinctive and easily recognizable. Specific determination of the majority of the scandent forms and of most monograptids is a matter for the specialist. Graptolite-bearing shales have an abnormally high content both of carbon and of iron pyrites. The origin of the carbon is uncertain, but it may be floating algae to which the graptolites were attached. The pyrite indicates that the waters immediately above the sea floor were deficient in oxygen and poisoned with hydrogen sulphide. A normal benthonic fauna, including many scavengers, could not survive in such foul bottom conditions. Anaerobic bacteria could, but they would only be capable of breaking down the soft parts, not the chitinous skeletons, of the planktonic graptolites which had sunk to the sea floor. The scarcity of graptolite fragments in coarser grained more sandy deposits must be due partly to greater agitation and current activity and partly to the abundance of scavengers able to live in the better oxygenated waters.

11 TRACE FOSSILS AND PSEUDO FOSSILS

The older geologists gave the name *Fucoid* to a wide variety of markings, which they had noted on the bedding planes of sedimentary rocks and interpreted as the remains of seaweeds. But in recent years, it has been apparent that 'fucoids' and other markings have been formed by a wide variety of organisms, other than seaweeds. These markings or **Trace fossils** have been studied with great thoroughness. Each morphological type recognized has been given a different name. In some cases it has been possible to relate a particular trace fossil to its marks. For instance, the five-rayed *Asteriacites* found in the Jurassic rocks of Germany record the resting place of starfish on the sea bottom, the shallow double rows of pits, named *Cruziana* were made by the walking legs of trilobites, for specimens of several genera of trilobites have been found at the end of *Cruziana* trails. More difficult to interpret is the great variety of tube-like structures, such as *Chondrites* (No. 150), that have been recorded. The tubes may be simple or U-shaped or they may be complexly intertwined. The majority of them were probably made by worms, either as a resting place or in process of working over an area of the sea floor when feeding, but it is extremely difficult to reconstruct their maker from simply the form and pattern of the tubes.

A further difficulty is that the structures produced in sediments by recent organisms have only been studied in a limited number of environments, mainly in coastal areas, including tidal flats. Again it is possible that some trace fossils were formed not at the surface of the sediment, but within the sediment itself, perhaps at the boundary between sand resting on a layer of clay. What is now preserved is the sand infilling of markings made on the surface of the clay.

A further difficulty is that these markings may be of inorganic and not of organic origin. As already mentioned in Chapter I, water may move easily through a sediment, especially whilst it is in an unconsolidated state soon after deposition. Salts may be dissolved in one place and then deposited in another, sometimes to produce concretionary bodies of bizarre shape. Concretions often form round organisms, for instance the flint enveloping the *Micraster* shown in No. 2 or the concretion from the Old Red Sandstone with a fish inside shown in No. 154. The familiar flints in the Chalk may on occasion have the most fantastic shapes, even simulating a human foot, a belemnite and so on. No. 151 from the Magnesian Limestone of Sunderland is not a compound coral, but a concretionary limestone. Such **pseudo fossils** do not show any trace of the cellular structure that is to be seen, though sometimes only under considerable magnification, in fossils, which have not suffered too much recrystallization or replacement. The shape may also give some guidance. Human remains are to be found in rocks of Pleistocene, not of Cretaceous, age. Admittedly one may expect to find belemnites in the Chalk, but their stout guards are resistant to change and when broken show a very different cross-section from a flint. But sometimes it is a difficult matter to determine whether a particular marking, concretion or structure in sedimentary rocks is inorganic or organic in origin. For instance, the Rhaetic Beds of the Bristol neighbourhood contain thin beds of limestone known as the 'Landscape Marble' because the patterns shown on polished specimens (No. 152) were thought to bear a resemblance to landscapes painted in the Japanese style. For many years the patterns were thought to be of inorganic origin, and only very recently has it been demonstrated that the 'hedge', the 'trees' and the 'canopy' are of algal (p. 188) origin.

It therefore follows that as well as the difficulty of recognizing fossils that are poorly preserved, one has always to be on the watch for trace fossils and pseudo fossils. The latter are misleading, the former may be informative, for there is always the possibility of finding a specimen which will enable a particular trace fossil to be related to its hitherto unknown originator.

12 VERTEBRATES

The **Vertebrates** (fish, amphibians, reptiles and mammals) are usually regarded as the most important group of the Animal Kingdom. In point of fact, they are outclassed by the arthropods, in numbers, variety of form and in the ecological niches occupied. The vertebrates are certainly the most important division of the phylum of the Chordata, for its other members, the hemichordates and the tunicates (sea squirts) are marine organisms only and of small size. In addition, from our standpoint, with the exception of the long extinct graptolites (p. 172), the hemichordates are extremely rare fossils, whilst fossil tunicates are unknown.

Many would regard the vertebrates as the most exciting of fossils. This is due to the resemblances to, and the differences from, the vertebrates familiar to us in our everyday life. Again, most people are familiar with the extinct vertebrates, often of giant size and bizarre shape, that are preserved in our museums and which may appear, sometimes in imaginary or greatly altered form, on the cinema or television screens.

It must be emphasized that vertebrate remains are not common fossils. Isolated teeth, the most durable of a vertebrate's hard parts, may be found in abundance at certain localities and one is always hopeful of coming across them in a variety of rocks, but skeletons, especially fairly complete skeletons, are very much rarer. Moreover, the collection of vertebrate remains, other than isolated teeth, is usually very much a matter for the expert, who is able to call on the specialists and the resources needed to deal with the masses of rocks, perhaps many tons in weight, in which the bones are embedded. Then the bones have to be cleaned of their matrix, or rather this has to be done as far as is possible without causing damage. Reconstructing the bones into their position in life is very much a matter for the specialist, for in most cases, not only has a three-dimensional animal been crushed flat on a two-dimensional bedding plane, but more likely than not the dead animal was not lying tidily on one side, but was perhaps half on its back with its limbs bent and its head at an unnatural angle, so that individual bones are not only broken, but are also considerably displaced.

The story of the fossil vertebrates is indeed a fascinating one, but it is as yet far from complete. Many of the links between the more fully documented parts of the story are based on a few specimens which have been painstakingly collected, prepared and studied by specialists. In the succeeding pages we shall sketch in outline the nature and development of the vertebrates. We shall concentrate on those vertebrate remains which are most likely to be found by the non-specialist. We would emphasize that irreparable damage can be done to vertebrate material, perhaps of great evolutionary importance, if it is incautiously handled. If one is fortunate enough to come upon vertebrate remains which look promising, it is far wiser not to attempt to extract them oneself, probably with inadequate tools, but to notify the nearest museum or university geology department, who will be able to advise as to the importance of the find and also as to the best methods of extraction.

Vertebrate Morphology

The distinctive features of vertebrate morphology are first a notochord or backbone, forming an axial stiffening which may be cartilaginous or a calcified vertebral column. The backbone provides support for the muscles, particularly for those required for the swimming or walking fins and legs. The nervous system is complex and is concentrated in the head with its highly developed brain. The respiratory system consists of gills in the lower and of lungs in the higher vertebrates. The body is clothed with a distinct skin, which may carry scales, feathers or hair to provide further protection.

A convenient classification of the vertebrates is into:

SUPERCLASS	CLASS	
Pisces	Agnatha, jawless	Silurian–Recent
	Placodermi, primitive jawed vertebrates	Silurian–Permian
	Chondrichthyes, sharks	Devonian–Recent
	Osteichthyes, bony fish	Devonian–Recent

SUPERCLASS	CLASS	
Tetrapoda	Amphibia	Carboniferous–Recent
	Reptilia	Permian–Recent
	Aves, birds	Jurassic–Recent
	Mammalia	Jurassic–Recent

The Chief Groups of Fossil Vertebrates

12*a* The **Agnathids** are represented today by the hagfish and the lampreys, parasitic forms attaching themselves to a fish by their sucker-like mouths and then feeding on its blood. In the highest Silurian and Devonian rocks occur the *Ostracoderms* (*bony-skinned*), of which the *pteraspids* and the *cephalaspids* are the best known groups. Both had a heavy head shield, which was flattened dorso-ventrally. In the cephalaspids the eyes were central, close together and looking upwards (No. 153), but in the pteraspids they were on the side of the head shield. Anteriorly the head shield of the cephalaspids was rounded, but in the pteraspids (Fig. 36) there was a long rostrum. The jawless mouth was on the undersurface of the head shield. Exceptionally well preserved specimens of cephalaspids from the Devonian rocks of Spitzbergen have shown that the brain was of primitive character, that there were ten gill openings and that the nerve canals were unusually large. Normally only the bony head shield is preserved. No trace of any internal skeleton has been found. If present, it must have been formed of cartilage and not of bone and therefore has not been preserved. The body of the pteraspids was covered by small scales, of the cephalaspids by large

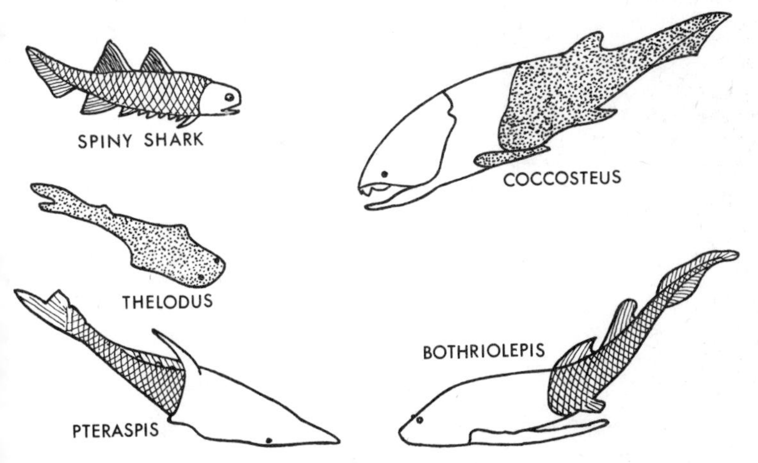

Fig. 36. *Agnathids (on left) and placoderms (on right). N.B. Heavily armoured parts are not shaded.*

rectangular plates. The pteraspids had no fins, though in the cephalaspids there were a pair of fins immediately behind the head shield. Both forms had powerful tails, especially the pteraspids. The pteraspids must have been quite efficient swimmers, the cephalaspids more sluggishly moving bottom dwellers. The dorsal spine and fins would prevent the bodies rolling from side to side. Other ostracoderms, such as *Thelodus*, are less well known, for they had no hard parts, though their body was covered with a great number of minute denticles, which tended to fall apart after death.

12*b* The **Placoderms** (*plated skinned*) are the most primitive of the *gnathastomata* (with jaw openings). They are restricted to the Palaeozoic Era and indeed are the only class of the vertebrates which is completely extinct. In the placoderms the jaws are of a primitive type, formed by modification of the anterior gill arches. The upper jaw was attached to the braincase and the lower jaw had a somewhat limited gape. Behind the jaws was a gill slit. In the more advanced fishes above the placoderm level the jaws are braced against a modified and strengthened hyoid arch, the hyomandibular, whilst the gill slit in front of it has been reduced to the spiracle. Paired fins were also present.

The placoderms are a varied group, including the 'spiny sharks' or *acanthodians*, with numerous fins supported by spines. Their body was covered with very many small diamond-shaped scales, whilst on their heads there were many small plates. The *arthrodires*, especially those of the higher beds of the Devonian were clearly efficient carnivores. The head was heavily armoured, but the body and tail were naked. An unusual feature was that the head shield was hinged on the ring of heavy shoulder or thoracic plates. Certain of the arthrodires were of very large size, reaching 30 feet in length. Much more bizarre in shape are the *antiarchs*, such as *Bothriolepis* (Fig. 36), bottom living forms with a heavily armoured head and body shield to which were attached long jointed appendages, which must have been used for movement across stream or lake bottoms. Indeed those of some specimens show very definite signs of wear by abrasion.

12*c* The **Chondrichthyes** include the *Elasmobranchs* or the sharks and rays. Unlike the other groups of Pisces they show little evolutionary change, for those of Upper Devonian times differ but slightly from the modern sharks, though the bottom living rays and skates did not appear till the Mesozoic. The distinctive features of the sharks are that the skeleton is cartilaginous, the articulation of the jaws is more complex than in the placoderms and they have great numbers of teeth, which develop in succession, are used and then drop out. As a result the teeth of sharks are by no means uncommon fossils. They are recognizable by their shape, but this includes not only biting teeth, but also the broad crushing teeth of the mollusc-eating bottom dwelling sharks and of the rays. A selection of elasmobranch teeth from different horizons are in Nos. 159–164 and this shows their characteristic colour and high lustre. Normally only isolated teeth are found, but occasionally one can come upon a number of teeth in their proper arrangement and then one can see that there are minor differences between the teeth in different parts of a jaw.

12*d* The **Osteichthyes**, the bony fish, are divided into two sub-classes, in part on the characters of their fins. In the *Actinopterygii* (the ray fins) the fins are supported by large numbers of parallel fin rays, but in the *Choanichthyes* (with lobe fins) there were bones along the median line of the fin with smaller bones or rays radiating from them. It was clearly from this *crossopterygian* type of fin that the walking legs of the tetrapods developed. But it is the actinopterygians that are the dominant group of the bony fish, especially in modern waters.

The main characteristics of the bony fish are first the perfection of their body form, fusiform and streamlined in those that live in open waters and need to swim fast, laterally compressed and deep bodied in the reef dwellers. Secondly the ossified brain case, whilst the skull is made up of a number of regularly arranged bones. The vertebral column is also ossified, there is a complex shoulder girdle, attached to the skull and to the pectoral fins. The primitive bony fish had functional lungs, but in the majority of the later forms the lungs had been modified into an air bladder, helping to control buoyancy. The body of the early forms was covered with thick heavy scales, but in later forms these became very much reduced in thickness.

The first bony fish are known from the Middle Devonian. The primitive actinopterygians or *chondrostei* include the *palaeoniscids*, most abundant in the Permian, and indeed represented today by a few forms such as the sturgeon. The living chondrostei show a great reduction of bony tissue both in the internal skeleton and in the outer covering of scales. Such loss of ossification with the passage of time is a feature that is shown in many other groups of fish, for instance the lampreys as compared with their heavily armoured Devonian ancestors, to a lesser extent in the sharks, and as we shall see later (p. 180) in the lung fish.

In Triassic times, more advanced actinopterygians, the *Holostei* or the ganoids, developed from the chondrostei and represented by a few living forms, such as the garpike of North American rivers. Whilst in the Jurassic appeared the first of the most advanced actinopterygians, the *teleosts*, a group which soon rose to dominance both in fresh and salt water.

But whilst the lobe fins are far less numerous and varied than are the ray fins, they are the important group from an evolutionary standpoint, for it was from the crossopterygian and not the actinopterygian type of skeleton that developed first the amphibians, and from them the reptiles and finally the mammals. We have already referred (p. 179) to the development of the tetrapod limb from the crossopterygian type of fin, but there are many other similarities, especially in the pattern of the bones that made up the skull. The choanichthyes are divided into two orders, the dipnoans, the lung fish, and the crossopterygians.

There is considerable resemblance but also significant differences, between the first actinopterygians and the earliest choanates, which are both found in rocks of Devonian age. The Devonian *lung fish* were fusiform and fairly active, but soon the group showed specialization to life in rivers reduced in the dry season to stagnant pools or even drying up completely. When this happens the lung fish burrow into the mud and survive for months, breathing air through an opening to the burrow. The Permian rocks of Texas have yielded the burrows of lung fish, some empty, showing that a successful escape had been made when the rains came, others containing a fossilized lung fish. From Devonian times the lung fish show several signs of degeneration. They have become rather eel-like in shape, with reduction in the size of the paired fins and the tail, there has been considerable reduction in the ossification of the skeleton, whilst the teeth instead of being borne on the margins of the jaws consist of large tooth-bearing plates. One might expect the lung fish to be the ancestors of the amphibians, but as has been said, they are only the 'collateral uncles'.

The *Crossopterygii* comprise two sub-orders, each very interesting, but for very different reasons. First the *Rhipidistia*, a progressive group from which the amphibians developed, secondly the *coelacanths*, an extremely unprogressive group which changed very little from the Devonian to the Cretaceous, when they were thought to pass into extinction. Then in 1938 a living coelacanth, *Latimeria*, was caught in a dredge off Madagascar. Since 1952 more have been caught. *Latimeria* is very similar in shape to *Macropoma* of the Cretaceous, so the coelacanths are an unusually stable group to be compared with the lingulids amongst the invertebrates (p. 133) in contrast to the great majority of the groups of fossil organisms, which show so much change with the passage of time.

12e The Amphibians

The Upper Devonian rocks of East Greenland have yielded a few specimens of a most important form, *Ichthyostega*, which shows a combination of fish-like and amphibian characteristics. It was at this time that air breathing fish must have struggled out on to the land, perhaps to seek fresh pools as their own dried up, perhaps in search of food. Life on land required considerable modification of the skeleton. In a fish the backbone is a lightly built structure (No. 165) for the body is largely supported by the water, but for a land dweller the vertebral column must be far stronger, firmly linked through the pectoral (shoulder) and pelvic (hip) girdles to the walking legs. The early amphibians show this development of stronger backbones and legs, with reduction in the size of the tail. By Upper Carboniferous and Permian times the *Labyrinthodont Amphibians* had developed, so called from the complex folding of the enamel layers shown in cross-section of their teeth, a feature also to be found in the crossopterygians. They were a varied group, some carnivorous, some herbivorous, some clearly only returned to the waters to breed, others may have spent much, if not all, of their time in the waters. But this period of amphibian supremacy was short lived, for by the end of the Triassic Period the labyrinthodonts had become extinct, and even before then the terrestrial forms had died out. The replacement of bone by cartilage in the latest labyrinthodonts clearly shows that they were water dwellers. The modern amphibians, the frogs and toads, branched off from the labyrinthodonts during the Carboniferous Period. They are a successful stock modified to live in or near water, feeding on insects and with their legs specialized to form a most efficient jumping apparatus. Their bones are thin and so they are rarely found fossil.

12*f* The Reptiles

A small tetrapod *Seymouria*, found in the Lower Permian beds of Texas, showing a remarkable combination of amphibian and reptilian characteristics, proves that the reptiles had developed from one branch of the labyrinthodonts. The major advance shown by the reptiles was that they laid eggs protected by a tough shell (cf. a turtle egg), so that they could become completely terrestrial without any need to return to the waters to breed. There were skeletal changes as well. The skull in many reptiles is deep instead of being flattened as in the amphibians, there are changes in the arrangement of the bones of the skull, most reptiles do not bear large teeth on their palate, the vertebrae are of different type, the shoulder girdle is better developed as are the limbs, whilst they have a horny, often scaley, skin.

The early reptiles of Permian and Triassic age were mainly quadripedal terrestrial forms, carnivorous and herbivorous, of relatively small size, but the Jurassic and Cretaceous beds have yielded an amazing variety of reptiles, often of gigantic size, including bipedal carnivores on land, swimming reptiles in the seas and yet other reptiles flying in the air (Fig. 37). This was the 'Age of Reptiles'. We shall refer here only briefly to some of the more important groups of the Mesozoic reptiles.

First the *Dinosaurs* (*terrible lizards*), a most varied group, including bipedal carnivores, some lightly built and active, others extremely massive, the largest terrestrial carnivores of all time, being up to 40 feet in length and probably six to eight tons in weight. Other dinosaurs, the largest terrestrial animals known, were the quadripedal swamp dwelling herbivores such as *Diplodocus*, 80 feet in length and weighing as much as 50 tons. There were also bipedal herbivores such as *Iguanodon*, swamp and lake dwelling 'duck-billed' dinosaurs, and a variety of heavily armoured quadrupeds.

The swimming reptiles comprise the *Ichthyosaurs*, with a beautifully streamlined body, long jaws wickedly armed with teeth (No. 167) and a powerful tail. In general form they resemble the modern porpoise. They were ovoviviparous, retaining the egg within the body until the embryo hatched out, for specimens of *Ichthyosaurus* with embryos in the body cavity have been found in the Lias of Wurtemburg. Specimens from Lyme Regis, preserving the stomach contents, show that ichthyosaurs fed on fish and on soft bodied dibranchiate cephalopods (p. 144). Belemnite guards have not been found, so either the ichthyosaurs, like certain modern whales, ate only naked cephalopods or their digestive juices were able to dissolve the densely crystalline belemnite guards or perhaps these indigestible portions were regurgitated.

Other marine reptiles include the long necked *plesiosaurs* with their limbs modified to form most effective paddles, and the *mosasaurs*, marine swimming lizards up to 30 feet in length. In the flying reptiles, the *pterosaurs*, the fourth finger of the fore limb was greatly elongated to support the wing membrane.

In dealing with the fish–amphibian–reptile transitions we have mentioned the discovery of connecting links. In the bony fish there is a complete sequence from the primitive chondrostei to the advanced teleosts, but we have little evidence of ancestral dinosaurs, ichthyosaurs, plesiosaurs or pterosaurs. They must have evolved during the Permian and early Triassic Periods, but marine beds of these systems are but scantily preserved on the present land masses, whilst many of the contemporary continental deposits are barren of fossils. There is a similar gap in the geological record at the close of Cretaceous times, when all these highly specialised groups of reptiles

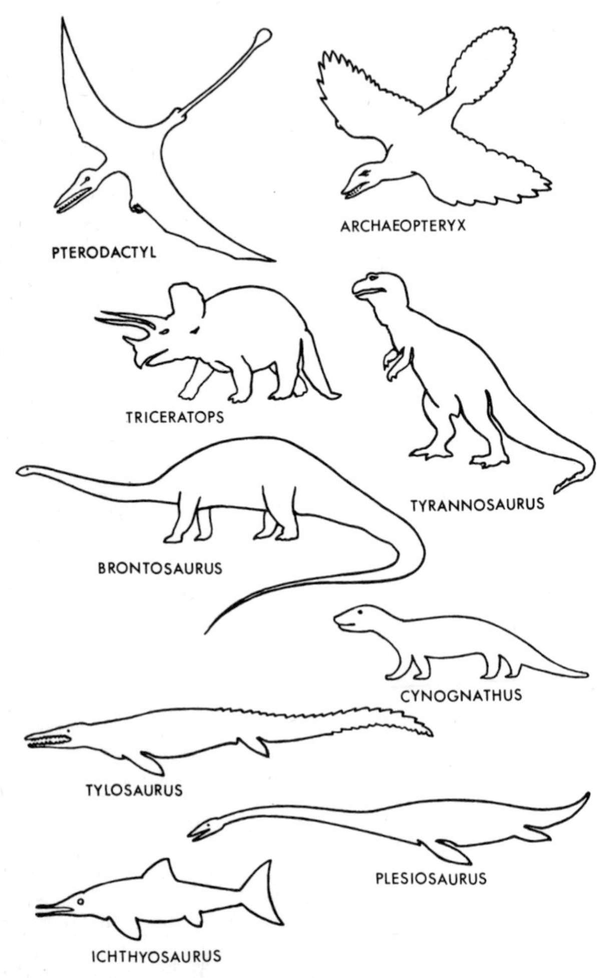

Fig. 37. *Representative Mesozoic reptiles. N.B. These reconstructions are not drawn to the same scale.*

became extinct. Many of the late Cretaceous reptiles were of giant size. Certain of the pterosaurs had a wing-span of over 20 feet, the plesiosaurs were up to 40 feet in length and we have mentioned the size reached by some of the dinosaurs. We do not know the causes for the comparatively sudden extinction of the Ruling Reptiles, but they may be related to the change in the food of the terrestrial herbivores caused by the rise of the angiosperms (p. 192), whilst the very extensive marine transgression of late Cretaceous times (p. 116) would have submerged the swamps which were the only possible habitat for such large and specialized form as *Diplodocus*. Climatic change and disease may have been other factors, but it is extremely difficult to get evidence of the latter from the fossil record.

But whilst such forms as these naturally excite our imagination, there is another group of reptiles of far greater evolutionary significance. These are the *synapsids* or mammal-like reptiles of late Carboniferous to late Triassic age. They are a group of quadripedal terrestrial forms (Fig. 41) ranging up to 8–10 feet in length. The early members are very definitely reptilian, the latter ones increasingly mammal-like, showing amongst other features, the development of the mammalian lower jaw which is formed of only one bone instead of several as in the true reptiles, the differentiation of the teeth from the closely similar teeth in the reptiles, to the markedly different incisors, canines and molar teeth of the true mammals, enlargement of the brain case and so on. It is impossible from fossil material to be sure whether or not some of the later synapsids were cold blooded as in the reptiles or warm blooded as in the mammals.

Certain groups of Mesozoic reptiles have survived into Tertiary and Recent times. An ancestral turtle is known from the Permian rocks of South Africa, true turtles from the Trias, the first crocodiles from the earliest Jurassic rocks and more advanced crocodiles from the Cretaceous. The late Cretaceous crocodiles show gigantism, for certain of them must have been at least twice as long as the largest living crocodile. The crocodiles, the closest living relatives to the dinosaurs, are another instance of a relatively unprogressive stock which has survived for very long periods. Another close relative to an important group of Triassic reptiles, is *Sphenodon*, the tuatara, which now survives under close protection in a few islands off New Zealand. Finally the lizards, ranging back to the Trias, and the snakes, which first appear in the Cretaceous but are very imperfectly known as fossils.

12g The Mammals

The first true mammals are known from a few isolated localities in rocks of Jurassic age, but in many instances only the teeth and jaws have been found. It is clear that during the Mesozoic period the mammals were not only very subordinate to the reptiles, but were also of small size, few being larger than the modern cat. The later Cretaceous rocks yield the first marsupials (carrying their young in a pouch) and also the first placental mammals, which fed on insects. Early in the Tertiary Era there was a great radiation of the mammals, comparable to the Reptilian Radiation of the early Mesozoic, with mammals occupying many of the ecological niches that had been vacated by the reptiles. In many groups of mammals one can recognize the same sequence. The relatively small brained and generalized archaic mammals of the early Tertiary give place to the later Tertiary modernized mammals with increase, sometimes marked increase, in the size and also potentialities of the brain, and with increasing specialization in the characters of the feet and of the teeth. The horses provide the most fully documented record, but comparable changes are shown in other groups such as the elephants, the biting cats and the primates. There is no space to deal with these here, but references are given in the bibliography.

12*b* The Aves

Finally the birds, which developed in the Jurassic from the reptiles. The Lithographic Limestone of Solenhofen has yielded two famous skeletons of the ancestral bird, *Archaeopteryx* (Fig. 37). One of the specimens is now in the British Museum, the other in the Berlin Museum. Both show many reptilian characteristics in the make up of the skull, the limbs, the long bony tail which is an extension of the vertebral column. But unlike the pterosaurs, they were feathered creatures with long flight feathers on the wings, on the body and also on either side of the tail. They could not only fly, but also must have been warm blooded, whilst the brain case showed that they had the complex nervous system needed by a flying animal.

More advanced birds, including diving birds, are known from the Cretaceous, but unfortunately our knowledge of fossil birds is very incomplete. For one thing, their bones are very thin and light, for many of them are filled with air.

The Geological History and Occurrence of the Vertebrates

The oldest vertebrate remains known are fragments of Ostracoderm plates found in a sandstone of Ordovician age in Colorado. Treatment with acid of British limestones of Silurian age has yielded the denticles of the agnathid *Thelodus* and fragments of placoderms. The earliest horizon to yield vertebrate remains in abundance is the *Ludlow Bone Bed*, the basal bed of the Devonian System in Shropshire. This is a condensed deposit, packed full of the denticles and spines of *Thelodus*. The overlying Old Red Sandstone is famous for the vertebrate remains which it has yielded. Pteraspids and cephalaspids in the Lower Old Red Sandstone of the Welsh Borderlands and the Midland Valley of Scotland, Oesel in Esthonia, Ringerike in Southern Norway; placoderms, crossopterygians and lung fish (dipnoans) in the Middle Old Red Sandstone of the Caithness Flags of north-east Scotland and of Russia; more advanced types in the Upper Old Red Sandstone of the British Isles, Wildungen and other localities in the Rhineland and the Baltic States. There are also many well known localities in North America, such as Scaumenac Bay in Quebec.

The so-called 'Fish Bands' of the Old Red Sandstone are often fossiliferous lenticles of very limited size, so that they may be worked out completely in the process of quarrying. This has happened in the case of the Dura Den Band from Fifeshire, which once yielded the superb slabs covered with specimens of the crossopterygian *Holoptychius* which are on view in the

British Museum (Natural History). Elsewhere the fish remains are disarticulated water sorted fragments found in the Bone Beds or in the Cornstones, concretionary limestones, of the Welsh Borderlands. Some of the early vertebrates that lived on the margins of the Old Red Sandstone land mass must have been trapped in bodies of water that dried up, others were swept down the rivers and came to rest when the currents slackened sufficiently. The widely scattered localities at which individual species have been found and the similarity between the vertebrate faunas of Europe and North America suggest that, like the modern eel, many of these early vertebrates must have spawned in marine waters, but spent most of their lives in the rivers and brackish water lagoons.

One is always hopeful of finding fish remains in the Old Red Sandstone. As indicated above they are not evenly distributed throughout the main mass of the beds, but are restricted to very definite horizons, often to particular bedding planes. Once one comes on fish remains, it is therefore wisest to work along that particular horizon. The remains may be mere impressions, they may be fairly complete (No. 154), they may be just isolated scales, but in each case they should be distinguishable from other fossils, particularly invertebrates, by their lustre and ornamentation.

In the marine rocks of Carboniferous age one may hope to find isolated teeth, including the broad crushing teeth and the spines (No. 155) of the sharks and 'spiny sharks'. The non-marine beds, especially the Coal Measures, yield in addition lung fish and crossopterygians, but no trace of the ostracoderms and antiarchs of the Devonian. These were already extinct. In addition, the Coal Measures have yielded a considerable variety of amphibians.

The Permo-Triassic beds of the Southern Hemisphere, particularly the *Karroo* Formation of South Africa, have provided a magnificent series of vertebrate faunas, comprising a wide range of amphibians and reptiles. These must have flourished in great abundance on the land mass known as *Gondwanaland*. But unfortunately the record from Europe is much more meagre. One well-known horizon underlying the Zechstein (p. 114) is the Kupferschiefer of Germany and the Marl Slate of England, a thin bed of a strongly bituminous shale or fissile siltstone, which contains many well preserved palaeoniscids. This concentration of fish remains used to be explained by rather sudden mass mortality in water containing an abnormal concentration of the salts of copper, but it is now appreciated that another factor may be of great importance. These beds are pyritic, their fauna is almost entirely nektonic with very little benthos. This suggests stagnant conditions on the bottom of a shallow sea, so that over a considerable period conditions were unusually favourable for preservation.

The New Red Sandstone of late Permian and early Triassic age near Lossiemouth in east Scotland contains amphibian and reptilian remains. These occur mainly as moulds, which have to be infilled with plaster of paris. Fissures in the Carboniferous Limestone of Glamorgan and the Mendips have yielded the skulls, teeth and long bones of the earliest mammals together with the remains of terrestrial dinosaurs. These fissures must have been open on a land surface of earliest Jurassic age.

The bones and skulls of the contemporary terrestrial vertebrates were washed into the fissures and later percolating water cemented the infill into hard breccia.

The marine Jurassic rocks of Europe, particularly the Lias of Lyme Regis in England and Wurtemburg in Germany and the Oxford Clay of Peterborough have yielded over the years to quarrymen and collectors a rich harvest of vertebrates, including many ganoid fish recognizable by their lustrous scales (No. 157), the vertebrae (No. 166), teeth, skulls (No. 167) and much more rarely, complete skeletons of the great swimming reptiles, the ichthyosaurs and plesiosaurs. Also the bones of pterosaurs, which may have damaged their wing membrane by misjudging their height above the sea surface. The Lithographic Stone is another famous horizon, for it is worked in great quarries and the quarrymen, are fully aware of the value of any finds. The Stonesfield Slate in the Oolites of Oxfordshire has also yielded many vertebrates, including fish (No. 162), dinosaurs and a very few mammalian teeth, but this bed of fissile sandy limestone is no longer worked to provide roofing 'slates'. The Purbeck Beds of Dorset have yielded a few mammalian teeth (from one small pocket), together with the remains of crocodiles and turtles as well as of fish.

The non-marine Wealden Beds of southern England yield ganoid fish, such as **Lepidotes** (No. 157) and more rarely dinosaurian remains. Towards the close of the last century, the workings of a coal mine at Bernissart, near Mons in Belgium, drove through the sides of a valley that had been cut into the Palaeozoic rocks and infilled with beds of Wealden age. The Wealden beds contained the well preserved remains of a herd of the herbivorous dinosaur, *Iguanodon*. These reptiles are now admirably displayed in the National Museum of Brussels. But most of the great dinosaurs, particularly the great swamp dwelling and the duck-billed dinosaurs, have been found not in Europe, but in North America, Asia and Africa. The teeth of fish, such as *Ptychodus* (No. 163) and *Corax* (No. 159), the bones of pterodactyls, the skulls and carapaces of water tortoises, are likely finds in the marine Cretaceous rocks, such as the Chalk, but the remains of the great swimming reptiles are much more uncommon than in the Jurassic rocks. The mosasaurs, the great swimming lizards, were named from being first discovered in the Upper Cretaceous rocks of the valley of the Meuse near Maastricht in Holland.

Most of our knowledge of the rise of the mammals in early Tertiary times comes from the sequence of richly fossiliferous horizons in the western United States. The European localities alone, few in number and widely scattered, would provide a very inadequate picture. For the Upper Tertiary, however, the European record is more complete. Vertebrate remains in Tertiary rocks of the British Isles are distinctly limited. The teeth of elasmobranchs such as *Odontaspis* (No. 159) are not uncommon at certain horizons in the Lower Tertiary beds of the London and Hampshire Basins. The Oligocene rocks of the Isle of Wight yield the remains, mainly of crocodiles, turtles (No. 168) and some mammals (No. 170). Over a hundred years ago a skull of *Hyracotherium* was found in the London Clay and recognized to be an ancestral type of horse.

Nearly a century later, when actual specimens and not drawings were compared for the first time, this skull was found to be identical with that of *Eohippus* (the dawn horse), thoroughly known from the many specimens that had been found in the Western United States. The pantotheres, a short-lived European side branch of the horses, are known from specimens found in the Isle of Wight and in the gypsum beds of Montmartre near Paris. After the Oligocene there is a long gap in the British record. The Bone Beds underlying the Crags of East Anglia have yielded numerous rolled teeth of vertebrates, a mixture of marine forms such as sharks (No. 160), whales and seals and terrestrial mammals including elephants, rhinoceri, etc. The gap in the British record is partly filled by European localities, such as the cave deposits of late Eocene to Oligocene age in Jurassic limestone at Quercy to the south of the Massif Central, the fissile limestone of Monte Bolca near Verona in Italy with its wealth of well-preserved teleosts (No. 158). Beautifully preserved fish are also to be found in the diatomaceous earths that infill many former lake basins in Germany.

Fluviatile deposits of Pleistocene age are always liable to yield the teeth and more rarely the skulls and bones of elephants (No. 169), rhinoceri, hippopotami (No. 171), the antlers of deer (No. 172), as well as the tools of man (No. 173). The famous Swanscombe skull, dating from about 250,000 years ago was found in the terrace gravels of the River Thames near Dartford in Kent. The first fragment was discovered in 1935. After twenty years further search, another fragment was obtained from the same seam of gravel, but 50 feet away. The parts fitted perfectly and clearly must have belonged to the same individual. Cave deposits may yield the remains of hyena, cave bear, etc., together with human tools and much more rarely human bones, such as those of *Homo neanderthalensis* first discovered at Neanderthal, near Dusseldorf in Germany. This side branch of the main human line lived between 70,000 and 40,000 years ago. The remains of various races of *Homo sapiens*, such as Cro Magnon Man and Grimaldi Man, that lived less than 20,000 years ago, have been found in many cave deposits of Upper Pleistocene Age in southern Europe. But we cannot do more here than to mention fossil man and his implements. Further information, especially of the discovery in Asia and Africa of early hominoids will be found in the works given in the Bibliography.

13 PLANTS

The study of fossil plants is really the province of the *palaeobotanist*, for the interpretation of well-preserved specimens requires a detailed knowledge of living plants. But to the geologist fossil plants, like any other group of organisms, are of interest, for they may provide valuable information as to the conditions that prevailed when the beds that contained them were laid down. In addition, if the sequence of past floras is known, then fossil plants may be of value in the dating of rocks.

A simple classification of plants is given on the following page:

Unicellular	Diatoms		Jurassic–Recent
Multicellular: Non-Vascular Plants	Thallophyta	(algae and fungi)	Pre-Cambrian–Recent
	Bryophyta	(liver worts and mosses)	Carboniferous–Recent
Multicellular: Vascular Plants	Pteridophytes	(club-mosses, horse-tails, ferns)	Devonian–Recent
	Spermaphytes	(seed bearing plants)	Carboniferous–Recent

The Morphology of Fossil Plants

13*a* The **Diatoms** are microscopic organisms, which are on the borderline between animals and plants. They float in the surface waters of seas and lakes as part of the plankton and therefore resemble animals in being mobile and free-living, but they resemble plants in performing photosynthesis. Their cell walls are impregnated with silica. In some modern Scottish lochs and in many Tertiary lake basins in Europe, where there was little or no other sediment, diatom skeletons have accumulated to form *diatomaceous earth*, an extremely fine-grained and inert substance, which is of great value for insulating and many other commercial purposes. Diatom remains (Fig. 15) have been recognized from as far back as the Cretaceous and they may well have a longer history.

13*b* The **Thallophytes** include the algae and the parasitic fungi. Unlike the higher plants, their bodies are not differentiated into root, stem and leaves. Many of the algae, such as the seaweeds, have only soft tissues and therefore are seldom fossilized in recognizable form. As we have already mentioned (p. 174) many of the 'fucoids' formerly supposed to have been the remains of seaweeds, have now been shown to have other origins. The *calcareous algae*, however, which secrete a hard limey skeleton, are important rock-formers. The framework of Recent and Tertiary coral reefs is built up largely by algae such as *Lithothamnion*, other algae play an important part in the reef-limestones of Jurassic age in southern Europe, whilst in the Carboniferous Limestone certain horizons are built up largely by algae such as *Solenopora*. Other algae become encrusted with calcium carbonate, precipitated from lime-saturated waters, forming the nodules or 'water biscuits', which are to be found in many modern lakes. *Girvanella*, *Mitcheldeania*, etc., which locally build up distinctive limestone bands in the Carboniferous Limestone have a similar structure. Blue-green algae colonizing tidal flats, bind sediment together with a mass of felted tubes, to form laminated *algal heads*, several inches in diameter (Fig. 174). Comparable laminated algal structures are to be seen at a number of horizons (e.g. No. 152) in the Phanerozoic rocks. The Pre-Cambrian rocks of Montana and parts of South Africa include limestones, built up of *stromatolites*, laminated structures, often of large size and of a wide variety of forms. In the past there has been much discussion as to whether these stromatolites were of inorganic or organic origin, but they are now generally accepted to be algal in nature. They also occur in the Pre-Cambrian rocks of northern Norway and have been reported from a few other European localities.

Another group of algae, the coccolithophoridae, secrete extremely minute discs and star-shaped bodies of calcite known as *coccoliths* (Fig. 15). They are marine planktonic organisms and on parts of the floor of the Atlantic Ocean, coccoliths are accumulating in sufficient abundance to form deposits of coccolith ooze. With the invention of the electron microscope, providing magnifications of twenty thousand times or more, it became possible to show that the familiar Chalk, particularly the finer textured parts of the Chalk, was made up very largely of coccoliths. Coccolith ooze is forming today in depths of one thousand fathoms or more, but the Chalk must not be regarded as a fossil deep sea ooze, for its macroscopic fossils and other evidence clearly show that the Chalk must have accumulated in a shallow sea whose depth probably rarely exceeded 100 fathoms. Conditions in the seas of late Cretaceous times must have been such that coccoliths were able to accumulate in much greater abundance and in shallower water than they can today. The grass-green algae include *Chara*, a bushy plant growing in modern lakes and ponds. It is a lime secreter, especially on its spherical spore sacs, which have a distinctive twisted spiral pattern. Such remains of *Chara* are not uncommon in freshwater limestones of Tertiary age.

We have already suggested (p. 188) that the diatoms combine certain features which are normally regarded as characteristic of the Animal Kingdom on the one hand, of the Plant Kingdom on the other. In recent years there has been an increasing tendency to recognize a third kingdom, that of the **Protista** (G2. *the very earliest*), to include all organisms that are unicellular and without a definite cellular arrangement. According to this definition, the Protista would comprise the Diatoms, the Thallophytes and the Protozoans.

13*c* The **Bryophytes**, the liver worts and the mosses, are traceable as far back as the Upper Devonian, but their structures are too fragile to be preserved except under very special conditions.

The Vascular Plants

These are highly organized with a root system for the abstraction of water and mineral salts from the soil, a stem formed of bundles of elongated cells through which the sap moves upwards and leaves which function for photosynthesis and transpiration. Reproduction is by spores in the Pteridophytes, by seeds in the Spermaphytes.

13*d* The **Pteridophytes** (vascular crytogams), the dominant group in the Upper Palaeozoic, are today very subordinate to the seed-bearing plants.

The Pteridophytes can be divided into:

Psilophytales	Devonian
Lycopodiales (club-mosses)	Devonian–Recent
Articulatales (horse-tails)	Devonian–Recent
Filicales (ferns)	Devonian–Recent

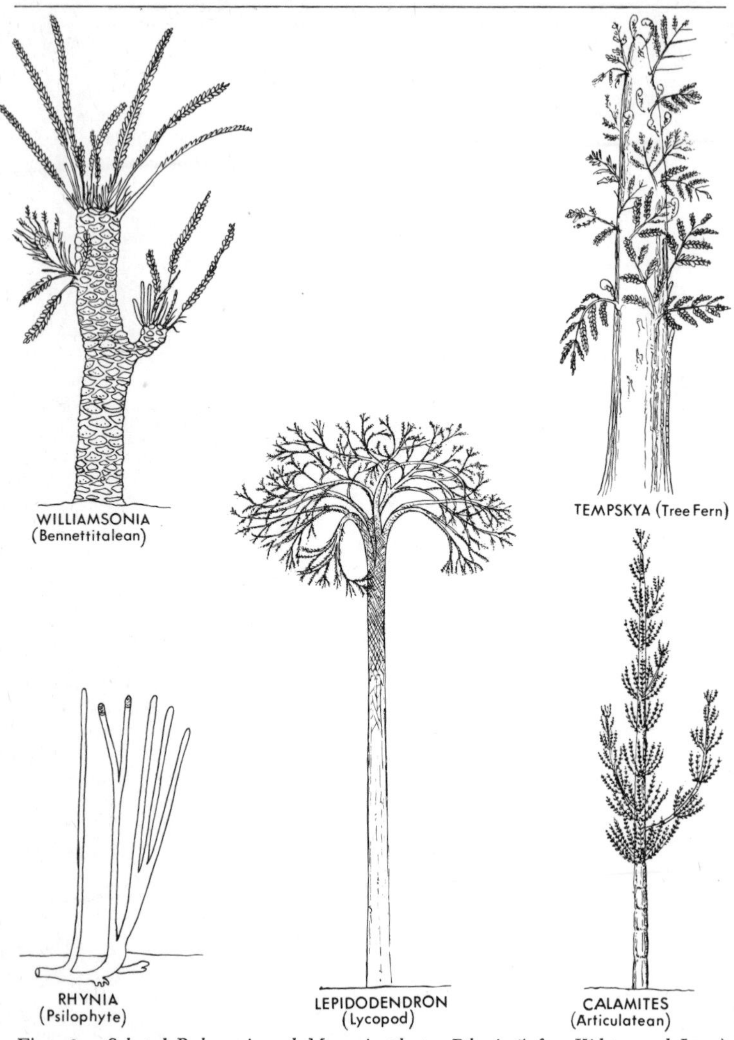

Fig. 38. *Selected Palaeozoic and Mesozoic plants. Rhynia (after Kidston and Lang), Lepidodendron (after Eggert), Calamites (after Dunbar), Tempskya (after Andrews and Keen), Williamsonia (after Sahni). N.B. These reconstructions are not drawn to the same scale.*

The *Psilophytales* are primitive land-plants showing the first stages of differentiation into roots, stem and leaves (Fig. 38). They are best known from the chert of Rhynie in Aberdeenshire of Devonian age.

The *Lycopodiales* are represented to-day by the club-mosses, such as *Lycopodium*, small creeping plants only a few inches in height, but in the forests of the Upper Palaeozoic the lycopods formed trees over 100 feet in height.

One difficulty with the nomenclature of fossil plants is that different names may be applied to the roots, the stem, the leaves and the reproductive organs of the same plant. This is because these were originally discovered, described and named as separate entities and it was only later that more complete specimens showed that, for example, *Stigmaria* (No. 176) were the roots of the stems known as *Lepidodendron* (No. 175), which bore leaves (*Lepidophyllum*) and reproductive cones (*Lepidostrobus*). On the other hand, one normally finds only disarticulated parts.

At a number of localities, for example, Victoria Park, Glasgow, *Fossil Forests* have been found with the stigmarian rootlets in the position of growth, spreading out along the bedding planes and radiating outwards from the broken-off stumps of the trees. *Stigmaria* consisted of four main roots diverging from the stem and then each main root forked repeatedly and gave off lateral rootlets. The stems of the Lycopods were unlike those of modern trees which are composed mainly of wood with only a thin covering of bark. The Lycopods had only a narrow axial stiffening of wood surrounded by a very thick layer of cortical tissue, so it has been suggested that the forests of the Coal Measure swamps could be blown down much more easily by the wind or laid low in other ways than modern forests. The stems of the Lycopods show a characteristic external pattern formed by the leaf-scars, which may be arranged as in *Lepidodendron* (No. 175) or in vertical rows as in *Sigillaria*. Specimens can sometimes be found with the short spiky leaves in position, but more commonly they have become detached from the stems and branches.

The *Articulatales* are represented today by the horse-tail *Equisetum*, a lover of wet ground where its stems, a few feet in height, arise from a horizontal *rhizome*. In the Coal Measure swamps, *Calamites* must have formed trees 50 feet or more in height. The stems are easily recognizable (Fig. 38) for they show a strong vertical ribbing, interrupted by *nodes*, from which arose the leaf-whorls known as *Annularia* (Fig. 39). The reproductive cones (*Calamostachys*, etc.) were carried at the tips of the main stems. *Sphenophyllum* (Fig. 39) with whorls of rather wedge-shaped leaves, is a climbing plant of Carboniferous to Lower Permian age, that can be grouped with the Articulatales.

The *Filicales*, the ferns, have very large leaves, bearing clusters of sporagia on their under surfaces, and but short stems. They are most important in Mesozoic and later rocks.

13*e* The **Spermaphyta** can be divided into:

Gymnosperms	Pteridospermeae	(Carboniferous–Jurassic)
	Cycadophyta	(Triassic–Recent)
	Ginkgoales	(Triassic–Recent)
	Coniferales	(Carboniferous–Recent)
Angiosperms		Cretaceous – Recent

The essential difference between the Gymnosperms (*naked seeds*) and the Angiosperms (*receptacle seed*), the flowering plants, is that in the former the seed is not completely enclosed in an ovary, whilst in the latter it is. The ovary is often very strong and resistant, as for example the many kinds of nuts or the hard seeds of sweet peas and other flowering plants, which the gardener is advised to scratch before planting, so as to assist the escape of the young plant.

The *Pteridosperms* were a very important part of the Coal Measure flora. A large number of leaf genera have been recognized, e.g. *Alethopteris* bearing at the terminal end of the axis of the leaf the seed *Trigonocarpus*, *Sphenopteris*, *Neuropteris* (Fig. 39), etc.

The *Cycadophyta* are stumpy palm-like trees with a crown of large leaves (Fig. 38). In the Cycads proper the seeds are borne in distinct cones, but in the Bennettitales in special flower-like shoots. The Mesozoic has often been referred to as the 'Age of Cycads', but within the last few years, the discovery of better preserved specimens has shown that the majority of the so-called cycads really belong to the Bennettitales.

Ginkgoales are today represented by *Ginkgo*, the maiden hair tree, with very distinctive wedge-shaped leaves (No. 177). *Ginkgo* is today only found native in parts of China and Japan, but in the Mesozoic the Ginkgoales had a world-wide distribution.

The *Conifers* (firs, pines, yews, etc.) have very small needle- or scale-like leaves and carry their sporangia in woody cones. They are the only group of gymnosperms that are abundant today, but even so they have been driven by the more successful angiosperms on to the areas of poorer soils or towards the tree-line of mountainous regions. In the Mesozoic, however, they were widespread.

The *Angiosperms* include the grasses, the flowering plants, the majority of the shrubs and of the trees that so beautify our gardens and the un-built up parts of the modern landscape. They are clearly the most highly organized group of plants with elaborate devices to ensure successful dispersal and propagation of their seeds. They are first known from the Lower Cretaceous rocks and by the beginning of the Tertiary Era had become the dominant group of plants.

The Geological History and Occurrence of Fossil Plants

The proven antiquity of Plants is much greater than that of Animals. The inarticulate brachiopod – trilobite fauna of the Lower Cambrian beds dates from about 600 million years ago. Traces of animals have been claimed from younger Pre-Cambrian rocks of somewhat greater antiquity, but there is still doubt as to whether these 'fossils' are really organic and if they are organic, whether they are animal or plant. But stromatolitic algal structures

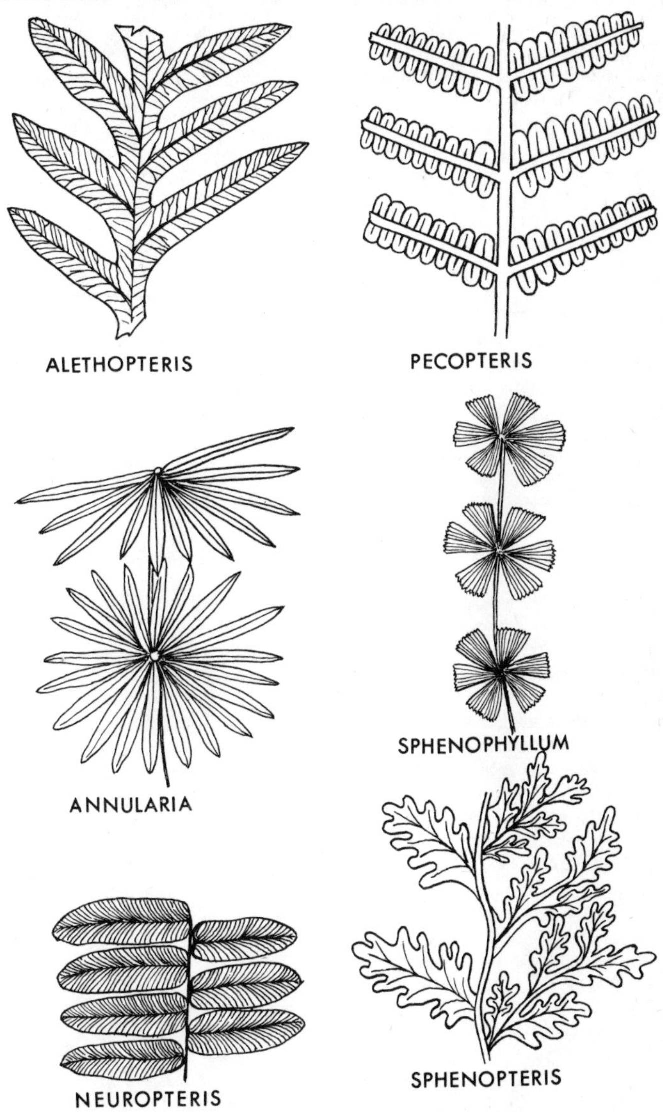

Fig. 39. *Leaves of common Coal Measure seed ferns.*

(p. 188) occur in the Pre-Cambrian rocks of Rhodesia, rocks which were deposited at least 3,000 million years ago, whilst in Ontario the fossilized spores and filaments of primitive fungi and algae have been preserved in chert in rocks whose age is at least 1,800 million years. Stromatolitic limestones are well developed in the younger Pre-Cambrian rocks of northern Norway and stromatolites have been recorded from other localities in the Pre-Cambrian rocks of Europe. We, therefore, have undoubted evidence of marine plants going back for some thousands of millions of years.

The colonization of the land by plants was much more recent. Spores of pteridophytic type have been claimed from the Lower Cambrian clays of Esthonia and fragments of vascular plants from the Middle Cambrian of Siberia. Undoubted psilophytes have been found in a number of widely scattered localities around the Silurian–Devonian boundary, but the geological evidence suggests that, as at Rhynie (p. 165), these plants were growing in swampy places.

The later Devonian rocks, however, have yielded a considerably more varied flora, which is directly ancestral to that of the Coal Measures. The coal-bearing Carboniferous rocks have long been famous for their plant remains, which may occur either as carbonaceous films or as impressions revealing only the outer form and surface markings or as petrifications or '*coal balls*'. In the coal balls, the internal structures of the plants are often preserved in exquisite detail in silica or in calcite, but as the concretions have usually formed round knots of vegetation, the fragments are often too small to give a good picture of the outer form. Stigmarian rootlets and the stem of lycopods are often preserved in the form of considerably flattened sand-filled internal moulds. The best place to search for plant remains is not in the actual coal seams themselves, but in the overlying shales, which are thrown aside to build up the unsightly tips around so many British mines. Here one can hope to find as well as roots, stems and branches of lycopods, such as *Stigmaria*, *Lepidodendron* and *Sigillaria*, the stems of *Calamites* and the leaves of many pteridosperms such as *Alethopteris*, *Pecopteris* (No. 178), etc. It has been shown that a sequence of successive floras can be recognized in the Coal Measures, but this is not the most practical way of zoning the beds, for it is a matter of finding specimens sufficiently well preserved to enable the species present to be identified. In the lower part of the Coal Measures it is usually easier to use the non-marine lamellibranchs (p. 151), but at higher horizons as they become less common increased use has to be made of plants.

The coal seams of Europe and North America originated as forest peats in low-lying lagoons and swamps. The climate must have been humid and equable, for the trees do not show annular rings, but not necessarily tropical. At times the peats must have accumulated to depths of scores, if not hundreds of feet, for the thicker coal seams are now several yards in thickness. It is estimated that the peat has been compacted by the weight of the thousands of feet of sediments that buried it in post-Carboniferous times to perhaps a fifteenth part of its original thickness. These periods of plant growth were ended rather abruptly by marine transgression submerging the swamps and then the cycle of the changing environments of the Coal Measures (p. 113) began again. But not all coals were formed in this way. The cannel coals are not underlain by seat earths as are the humic coals. They are composed of plant debris that drifted or floated down the rivers and then sank in patches of slack water to accumulate as organic sediments mixed with considerable amounts of clay and silt. In the same manner fragments of the coniferous vegetation that lived on the higher ground behind the swamps are to be found in the sediments that separate the coal seams.

Towards the close of the Carboniferous Period there was a world-wide change to more arid conditions, so the swamp-loving plants quickly diminished in importance. The flora of the Permian Period is not too well known, for fossiliferous localities are few, but it was clearly directly related to that of the underlying Carboniferous beds. During the Triassic Period, however, there were great changes, with the rise of the cycadaceous plants; conifers such as *Voltzia* and ferns persisted, together with horsetails, but the pteridosperms were passing into extinction. The Estuarine series of the Midland Jurassic, the Purbeck Beds and the Wealden Beds have all yielded rich floras. The well known 'Fossil Forest' at Lulworth on the Dorset coast shows the stumps of coniferous trees underlain by a fossil soil or Dirt Bed. The trees are silicified, but they are heavily sheathed in secondary tufa and with the centres of the stumps often hollowed out, the old quarrymen's terms of 'birds nests' or 'burrs' are at least descriptive (page 47). The Estuarine Series on the East Yorkshire coast sections and the Wealden Beds at Hastings yield the stems of the horsetail *Equisetites* (No. 179), the leaf impressions of a wide variety of ferns such as *Onychiopsis* (No. 10) and stems such as *Tempskya* (No. 180), the stems, leaves and reproductive organs of cycads and bennettitales, and fragments of coniferous wood. Soil beds full of the rhizomes and broken stems of the horsetail *Equisetites* have been traced throughout much of the Weald of Kent and Sussex. They must represent extensive reed swamps growing in only a foot or two of water.

The earliest *Angiosperms* are claimed from rocks of Lower Cretaceous age. By the close of the Cretaceous Period the vegetation was essentially of modern type. Drifted fragments of coniferous wood are sometimes to be found in the Chalk and other marine Upper Cretaceous rocks, but owing to the extensive marine transgression which submerged most of Europe, our knowledge of the Upper Cretaceous floras has to come from richly fossiliferous extra-European localities in Greenland, the western United States and Japan.

The leaves of *Angiosperms* are beautifully preserved in the continental facies of the Tertiary beds either in pipe clays, such as the Leaf Bed at Alum Bay in the Isle of Wight (No. 183) or in diatomaceous earths in the lake basins of Germany. One particularly rich horizon of Miocene age is at Oeningen on Lake Constance on the German–Swiss border. Drifted plant remains may be found in the marine beds, not only pieces of wood, but also cones and seeds, such as those of the tropical palm *Nipa* from the London Clay of Sheppey (No. 181). Periods of plant growth, long enough for the formation of peat, produced beds of Brown Coal or Lignite. In the English

Tertiary rocks, the seams of lignite are too thin to be worth working, but in many places on the continent of Europe, the brown coals are a most important source of fuel. For example in the Miocene basin to the north of Cologne, the main seam reaching a thickness of over 300 feet, is worked in gigantic opencast pits. Important constituents of the peat are the conifer *Sequoia*, now restricted in its native state to a belt along the Pacific coast of North America, where one species produces the well known 'big trees' of California, and also a species of *Taxodium*, closely allied to the swamp cypress of North America.

Plant remains are also to be found in the Tertiary rocks of the Volcanic Province (p. 117). The moulds of tree trunks or charred trunks which have been partly replaced by basalt, are known from lava flows in Mull and Antrim, whilst the fossil soils between the lava flows have yielded a varied flora. The Oligocene amber of the Baltic coast encloses not only insects, but also fragments of leaves and flowers.

The flora of the British Tertiary Beds is tropical or sub-tropical in character. We have already mentioned the Malayan palm *Nipa* from the London Clay, but more complete evidence is provided from the inter-volcanic horizons. The species of *Quercus* (oak), *Platanus* (plane), *Corylus* (hazel), *Ginkgo* (maidenhair tree), etc., found there are those now living today in southern China. Moreover, essentially the same flora is not only found throughout Europe, but also as far northwards as Spitzbergen and northern Canada. The palaeobotanical evidence strongly suggests that in Lower Tertiary times, at least, climatic conditions throughout the Northern Hemisphere were essentially of a uniform subtropical nature without the marked latitudinal variation of the present. But whilst the Tertiary plants are valuable in this sense, they are not nearly so helpful for dating purposes, for there is too little variation in the floras with the passage of time.

Towards the close of the Tertiary Era, however, these widespread genial conditions came to an end, with the development of continental ice sheets in northern latitudes. To the south of the ice sheets stretched barren tundra with a vegetation of mosses, lichens, dwarf birch (*Betula nana*) and arctic willow (*Salix polaris*). Towards the continental interior the tundra graded southwards into grassy steppes, but in more maritime climates the tundra was fringed by coniferous forests (birch and pine) and then further southwards the conifers gave way to mixed deciduous forests with oak predominating. During the climatic oscillations of the Pleistocene with its sequence of interglacial and glacial periods, these floral belts swung northwards and southwards across Europe. As we have already mentioned (p. 117) the deposits laid down during each interglacial episode were largely ploughed off by the next advance of the ice sheets. But locally they have been preserved mainly in sheltered lake basins or in the deposits in river valleys and estuaries. Much of the evidence can only be gained by sinking borings.

Floral evidence is proving of great value in unravelling the complex story of Pleistocene and Holocene times, but it is a special kind of floral evidence. **Palynology** or the study of the spores and pollen of plants, first developed in

Sweden in the study of the sequence of floral zones recognizable in the deposits laid down during the last retreat of the ice sheets. The spores and pollen of plants are widely distributed by the wind and other causes. Some of the material will settle in lakes, ponds and swamps, and become part of the sediment that is accumulating there. With suitable techniques the spores and pollen can be isolated, studied under the microscope, identified and counted. The percentage frequency of the different types of tree-pollen present at any horizon gives a picture of the vegetation at that time. In the early studies only tree-pollen was counted, but in the more recent work the whole pollen-spectrum including grasses, flowering plants, etc., is considered. Throughout much of north-western Europe, including the British Isles, the following palynological sequence can be recognized:

Present to 500 B.C.	Sub-Atlantic	Alder–oak–elm–beech
	Sub-Boreal	Alder–oak–elm–lime
3,000 B.C.		
	Atlantic	Climatic optimum
6,000 B.C.		
	Boreal	Pine–hazel
7,000 B.C.		
	Pre-Boreal	Birch–pine
8,000 B.C.		
	Upper Dryas	Tundra
9,000 B.C.		
	Allerød oscillation	Birch woods
10,000 B.C.		
	Lower Dryas	Tundra

The Allerød oscillation, a milder episode, between the two tundra phases named after *Dryas octopetalata*, is a widely traceable horizon. The dividing line between Post Glacial (Holocene) and Late-Glacial (highest Pleistocene) is drawn between the Upper Dryas and the Pre-Boreal assemblages.

In recent years it has become clear that palynology can be of great value at earlier horizons. A pollen-bearing interglacial deposit will show the sequence tundra–mixed oak–tundra. Furthermore the pollen-spectrum of each of the different interglacial deposits has its own distinctive features, not so much in the different types of trees represented, but in their relative abundance and also order in which they become abundant. Palynology is proving, in the hands of the expert, a more valuable tool for the correlation of Pleistocene deposits than is the study of their vertebrate and invertebrate fossils. For one thing pollen-bearing deposits are much more common than are those containing reasonably well-preserved faunas.

But palynological work is not restricted to the Pleistocene. Pollen and spores are another variety of microfossils and the micropalaeontologists of the oil industry are using them as such. Certain Coal Measure geologists are

V. THE COLLECTION AND EXTRACTION OF FOSSILS

EQUIPMENT

A hammer, a chisel, a strong-bladed knife and wrapping materials form the simple basic equipment needed for fossil collecting, but they must be chosen with care. Whilst any good ironmonger can supply cold chisels and strong knives or trowels, he is very unlikely to have safe and suitable hammers in stock. The steel of ordinary domestic and many trade hammers is far too soft for use on most rocks. It will splinter and flying fragments of steel can cause nasty wounds, particularly to the eyes, not only of the hammerer but also of his companions.

A geological hammer as supplied by the Cutrock Engineering Company, 35 Ballards Lane, London, N.3., or by Messrs. Gregory Botley, 30 Old Church Street, Chelsea, London, S.E.3., is a carefully designed tool. The head, made of specially hardened steel, is firmly attached to the shaft. Hammers with the head and shaft cast in the one piece are certainly very strong, but they are short and do not have the well-balanced feel that can be obtained with a wooden shaft of the right length. But the wood may shrink, loosening the head, though this can usually be quickly put right by soaking the hammer head in a rock pool or a stream. In dry climates, however, shrinkage can become such a problem that a steel shafted hammer is a necessity. The head of a geological hammer is square at one end with chamfered edges, whilst the other end either tapers to a point or has a straight edge, usually at right angles to the shaft, though some prefer the chisel edge to be parallel to the shaft. A convenient general purpose weight for a geological hammer is between 2 and 3 pounds, but for breaking very hard rocks one may need a 7- or 14-pound head with a corresponding longer and thicker shaft, whilst for the delicate work of trimming surplus matrix off specimens one requires a hammer weighing only a few ounces. Similarly one needs a range of cold chisels ranging from a 'baby', a few inches in length with a sharp edge of a quarter of an inch, to heavy chisels a foot or more in length and an inch or more in width. If one knows the type of rocks on which one will be working, then one can select tools of the appropriate weight and avoid carrying a lot of unnecessary ironmongery.

Hammers and chisels are, however, not the best of tools when working in unconsolidated deposits, such as sand and clay. A broad bladed entrenching tool, obtainable from suppliers of camping equipment, is far more effective for cleaning up a face; whilst stout knives or flat-bladed trowels can be used for digging out the fossils, aided perhaps by brushes, if the fossils are unusually fragile.

As in most sports and hobbies, it is worth spending care on one's equipment, making sure that it feels right and then learning to use it to the best advantage. There is quite an art in using a hammer properly. Instead of just pounding on the rocks, which is liable to produce only dust, one uses either the chisel edge or one of the corners of the square end to strike carefully placed blows, which will split the rock along the bedding planes where fossils are most likely to lie, or in a more massive rock will open cracks so that a block of rock can be detached cleanly. If necessary, the wedging off of blocks or the opening of bedding planes can be helped by driving in carefully placed cold chisels.

Different types of rock break differently and it is as well for the beginner to put in a little practice, getting the 'feel' of a particular rock type before he starts hammering out fossils. It is far too easy to ruin a good specimen by one ill-placed or too strong blow of a hammer.

Small fossils in unconsolidated deposits can also be collected by using sieves of suitable aperture. One practical difficulty in the field is that the sieves so easily become clogged, especially if the material being sieved is clayey and of high moisture content. Under dry weather conditions, however, the more durable fossils such as the bones and teeth of fish can be quickly sieved out of sands. If there is plenty of water available, as on a beach or by a stream, then the material can be sieved wet and the matrix gradually washed away, but shaking in a sieve is always liable to damage or destroy the more fragile fossils.

In some beds, particularly in the unconsolidated sands and clays of Tertiary and Pleistocene age, the fossils may be so fragile that they need hardening in the field. Alternatively a block of matrix with the fossil in it must be dug out, wrapped, and the whole brought home for treatment indoors, but obviously if this is done, the weight problem may become acute. A useful and easily obtainable hardening medium is Rawlplug Durofix diluted in equal parts of amyl acetate and acetone. It is inflammable and leaves a glossy surface on the fossil.

Having collected the fossils it is essential to protect them adequately from damage whilst they are being transported home. Every fossil or lump of rock containing fossils must be wrapped either in newspaper or placed in bags. Polythene bags are now replacing the older fashioned cloth or canvas bags. The more fragile fossils need the greater protection of a tin or box lined with cotton wool or, if this has been forgotten, moss or anything else soft that is to hand.

If large blocks containing fossils have been collected, it is advisable to remove unwanted matrix before moving on. This can be done by light taps with a trimming hammer, but again there is always the risk of ruining the specimen. If in any doubt, it is far safer to shoulder a little more weight than to attempt cleaning up operations which can be done far more safely indoors. A rucksack is the most convenient way of carrying one's tools and specimens.

Finally, before moving on to a new collecting area, all specimens must be labelled with the locality and if possible the horizon from which they came. The safest way is to number each specimen, and write the full details against a number in a field notebook. The number can be written on large specimens or a numbered ticket included with the wrappings of smaller specimens. If this is not done and the specimens are not unpacked for some time after they have been collected, then there is a serious danger of errors being made in the localities from which the specimens came. This might have serious consequences if the specimens were of considerable interest for zonal or other scientific purposes. An unlocalized, or worse still an inaccurately localized, fossil is of greatly reduced value and may indeed be seriously misleading.

A hand lens of 8–10 times magnification is often invaluable in the field both in looking at the detail of fossils and also in deciding whether small fragments of possible fossils in the rock are worth further attention.

COLLECTING FOSSILS

One is often asked 'where can I go to find fossils?' Not an easy question to answer, for it involves two things. First to select an area where fossils are to be found and secondly advice on how to find them on the ground. As we have seen in the preceding chapters fossils occur in certain beds of the sedimentary rocks; they are rare, and usually so distorted that they are difficult to recognize, in rocks that have suffered metamorphism and except in some beds between lava flows, are not to be found in areas of igneous rocks. But it is only certain sedimentary rocks that yield fossils in abundance and even in these the fossils may be restricted to particular strata. Sometimes one has to find a layer only an inch or so in thickness in a succession of several hundred feet of rocks. On the other hand, there are localities, such as the Dorset–Devon coast on either side of Lyme Regis, where fossils are so abundant and well preserved through a varied range of sedimentary rocks that anywhere along this coast one has a reasonable chance of good collecting.

There is no one book which answers completely satisfactorily the question above. In 1954 the Palaeontolographical Society published *A Directory of British Fossiliferous Localities*, but unfortunately this is not as comprehensive as the title suggests nor are the directions as precise as one would like. The original edition has recently been reprinted. The Geologists' Association are producing a series of Guides, each of which contains a number of geological itineraries in a particular area. Full details of fossiliferous localities are given in many itineraries, but certain of the guides are concerned with areas of igneous or strongly metamorphosed rocks. These Guides can be obtained for between 20p and 40p each from 'The Scientific Anglian', 30A St. Benedict's Street, Norwich, NOR 24J. The Association also arranges a programme of field meetings often to fossiliferous areas, both in Great Britain and abroad. Membership of the Association (Secretary Dr. F. H. Moore, 278 Fir Tree Road, Epsom Downs, Surrey) is open to anyone interested in Geology. The Institute of Geological Sciences publish at very moderate cost a series of British Regional Geologies. These are general accounts of the geology of a particular region (Wealden District, Northern Highlands, etc.) and whilst the fossil content of the sedimentary rocks are mentioned and often illustrated, fossiliferous localities are usually not described in detail. These Regional Guides do, however, provide a good geological background for any particular region. The Geological Survey have published 1 inch to the mile geological maps of a large part, but not the whole, of Great Britain. For many of these maps, Sheet Memoirs are available. These contain exhaustive accounts of the geology of the area ground and usually fossiliferous localities are described in detail. A large number of these Memoirs were, however, written many years ago, so that the sections mentioned may have been filled in or built over. Anyone can consult the Survey maps and memoirs in the Library of the Geological Museum, Exhibition Road, South Kensington, London, S.W.7. This Museum and the adjacent British Museum (Natural History) are well worth visiting, for they contain well arranged exhibits of fossils.

Below are listed a number of other recent books which contain useful guidance to places where fossils may be found:

Charlesworth, J. K. *Historical Geology of Ireland*, 1963. Oliver and Boyd. 565 pp.

Davies, G. M. *The Dorset Coast. A Geological Guide*, 1964. Black. 128 pp.

MacGregor, A. R. *Fife and Angus geology – an excursion guide*, 1968. Blackwood. 226 pp.

Mitchell, G. H. *Edinburgh Geology. An Excursion Guide*, 1960. Oliver and Boyd. 222 pp.

Neves, R. and Downie, C. *Geological Excursions in the Sheffield Region and the the Peak District*, 1967. University of Sheffield. 163 pp.

Sylvester-Bradley, P. C. and Ford, T. D. *The Geology of the East Midlands*, 1968. Leicester University Press. 400 pp.

Versey, H. C. *The Geology of the Appleby District*, 1964. Whitehead. 44 pp.

In the following parts of the *Welsh Geological Quarterly* published by the South Wales Group, Geologists' Association are annotated bibliographies and indexes of geological excursion guides to A. Scotland, 1967, vol. 2, No. 3, B. Wales and the Welsh Borders, 1967, vol. 3, No. 1 and C. England, 1968, vol. 3, No. 3 and 4.

The Yorkshire Geological Society (Librarian, Department of Geology, The University, Leeds, 2) have published useful low-priced guides to the following areas: Ingleborough, Alnwick (Northumberland), Cliviger Valley, near Burnley, Lancashire, and the area between Market Weighton and the Humber, whilst the *Geology of Norfolk*, issued by the Paramoudra Club now the Geological Society of Norfolk and obtainable from the Castle Museum, Norwich, gives details of many fossiliferous localities, especially in the Chalk.

There is always the possibility of finding new fossiliferous localities. If one knows from Geological Survey maps or in other ways, the probable run of beds likely to be fossiliferous, it is well worth while examining any trenches, new road cuts or other excavations, which may cross them. But with modern methods of excavation and filling, such temporary exposures are unlikely to remain open for long.

Needless to say permission should be obtained to enter land that is not in examined to see how much further work on them is necessary. Pyritized fossils, especially those collected on a sea shore, are very prone to decomposition. If they are left exposed to the air, the pyrites, if it is in the unstable form of marcasite, will oxidize to ferrous sulphate and in a few years all that will be left will be a heap of whitish powder. Such fossils can be protected by dipping them in hot candle wax. Do not let the wax boil, drain off any excess or wipe it off with carbon tetrachloride, easily obtainable from ironmongers, handicraft shops etc. under the name of Thawpit.

Most fossils, when brought back from the field, have matrix adhering to them and this must be removed. Depending on the nature of the matrix, it can be washed off, brushed off or, if it is harder, careful work with a needle or fine knife will be needed. It is most important not to scrape the matrix away from the fossil as this will only mark its surface. If one presses the tool carefully down on to the matrix at right angles to the surface of the fossil, then pieces of the matrix will break off cleanly without damaging the fossil. In the same way one can often 'jump' fossils clean out of a rock with well-struck blows of a hammer. The development of fossils requires much skill and patience and it takes a little time to develop one's technique. A Burgess Drill, obtainable for a few pounds from a handycraft shop or a dentist's second-hand drill, even of the now obsolete treadle-driven type, are extremely useful, for they are versatile and enable one to work much faster than with hand tools alone. Sometimes fossils get broken either in the field or during the cleaning process and then it is necessary to use an adhesive. Uhu, a synthetic resin, is very useful for mending small pieces, for it sets very quickly and is not flexible. Araldite, sold in handyman shops, is another good adhesive for this purpose, but it takes about two days to harden. If the fossils occur as moulds, then it is necessary to take casts, having first dusted the surface with French Chalk or soap solution to act as a releasing agent. Ordinary Plasticine can be used, for it takes a good impression, but it is not flexible and therefore may become distorted whilst easing it out of cavities. The alginate impression compounds used by dentists, such as Tissutex, obtainable from the Dental Manufacturing Company, are flexible and show fine detail. Another useful medium is Vinagel Putty, sold by P. K. Dutt & Co. Ltd. This has to be heated for 20 minutes in air at 120–170°C to harden and remains slightly plastic.

If one finds silicified or phosphatized fossils in limestone, then it is possible to dissolve away the matrix with dilute hydrochloric or acetic acid. But such treatment requires careful watching, for if the fossil is only partially silicified, it may be attacked more rapidly than the matrix.

The detailed structures of fossils can often be revealed by a cut and polished surface (e.g. Nos. 28, 31, 34, 83 and 111). If a lapidary's or high speed diamond impregnated saw is available, the work of cutting is greatly facilitated. If not, a smooth surface can be ground across the specimen by hand rubbing on thick glass plates using carborundum powder as an abrasive and water as a lubricant. Start with 120 grade and then work down in about six steps of successively finer grades, finishing with 600 grade. It is most important at every change to a finer grade to make sure that the specimen and the plates have been washed clean of the coarser material, otherwise even a single grain will produce disfiguring scratches. If a power-driven drill is available, the polishing can be done with buffing pads using jeweller's rouge or Aloxite. If the polishing has to be done by hand, use jeweller's rouge on felt stretched over glass plates. To obtain the best finish lacquer the polished surface with a clear lacquer. The chrome protectors supplied by Holts for cars, are quite satisfactory.

There is much scope for working out one's own techniques for the preparation of fossils; so much depends on the type of preservation, the nature of the matrix, and the particular morphological features that need to be emphasized. Much can be done by the skilful adaptation of tools or materials designed for other purposes.

Brief reference has been made in Chapter IV to certain techniques used in the study of microfossils (p. 119), bryozoans (p. 130), brachiopods (p. 138) and vertebrates (p. 176). Kummel and Raup's Handbook (see Bibliography) gives an exhaustive account with full bibliography of techniques, but is expensive, often repetitive and many of the techniques require elaborate apparatus. Much simpler, but clear guidance is to be found in the British Museum (Natural History) Handbook. *Instructions for Collectors, Fossils, Minerals and Rocks*, whilst Chapter XII of the latest edition of Morley Davies's *An Introduction to Palaeontology* is helpful and concise with useful references.

A. J. Rundle has published in *The Mercian Geologist*, Vol. 2, 1968, a very helpful article on some of the simpler and cheaper methods for the collecting and preserving of fossils from unconsolidated deposits.

An excellent account by A. C. Higgins and E. G. Spinner of the techniques used in the extraction of microfossils has been published both in *Geology*, 1969, vol. 1, the journal of the Association of Teachers of Geology and also in the *Welsh Geological Quarterly*, 1968, vol. 4, no. 1. Copies can be obtained from the National Museum of Wales, Cardiff.

The catalogues of the firms listed below are worth consulting, for many of the tools, storage equipment supplied for other branches of natural history can be used equally well for palaeontological purposes.

P. K. Dutt & Co. Ltd., Clan Works, Howard Road, Bromley, Kent.

Watkins and Doncaster, 110 Parkview Road, Welling, Kent.

World Wide Butterflies, Over Compton, Sherborne, Dorset.

Gaps in one's collection of fossils can be filled, for the time being, by buying from firms such as S. A. Baldwin, 32 Highfield Road, Purley, Surrey or Ammonite Ltd., Llandow Industrial Estate, Cowbridge, Glamorgan, who specialize in producing excellent replicas of fossils. The originals are often perfect specimens housed in one of the national museums. Whilst these casts are usually much more durable than a real fossil, they are considerably lighter in weight and are usually uncoloured. Prices range from about 10p a specimen upwards, so it is advisable to write for a price list before ordering.

VI. THE IDENTIFICATION OF FOSSILS

NOMENCLATURE

The scientific names of animals and plants, in contrast to their popular or vernacular names, are for international usage and therefore scientific nomenclature has to be governed by a strict code of rules. The foundations of the system in use were laid by the Swedish botanist Linnaeus (1707–1778). He divided the organic world into two Kingdoms, the Plant Kingdom and the Animal Kingdom. The majority of organisms can be assigned without difficulty to one or other Kingdom, but as we have mentioned (p. 189), some would now recognize a third Kingdom, the Protista, to include those organisms, such as the diatoms, which combine characters found only in plants with others typical of animals.

Each Kingdom is divided into major groups or phyla. Phyla are subdivided into classes, classes into orders, orders into genera and genera into species, the smallest classification unit. Man, for example, belongs to the species *Sapiens* of the genus *Homo* of the order Primates of the class Mammalia of the subphylum Vertebrata of the phylum Chordata. In practice one normally states only the generic and specific name of an organism e.g. *Homo sapiens*. By the International Rules of Nomenclature, the generic name is a Latin noun or a word treated such. It is written first with an initial capital letter followed by the specific name with a small initial letter. So that generic and specific names can be clearly recognized as such, it is the convention to print them in different type from the remainder of the text, usually in italics.

Systematics or taxonomy are alternative names for the particular branch of science which is concerned with the classification and nomenclature of organisms. The taxonomic palaeontologist has to work within a complex and rigid framework of rules, which we cannot consider in detail here. Anyone interested is referred to Chapter XIII of the latest edition of Morley Davies's *An Introduction to Palaeontology*, for this gives clear and succinct guidance.

THE SPECIES PROBLEM

With living material, the members of a species should not only have a great number of morphological characteristics in common, but they should also be capable of interbreeding. Within the species there will be considerable variation in minor features which are not regarded as of classificatory importance – the type of variation that one sees in any gathering of

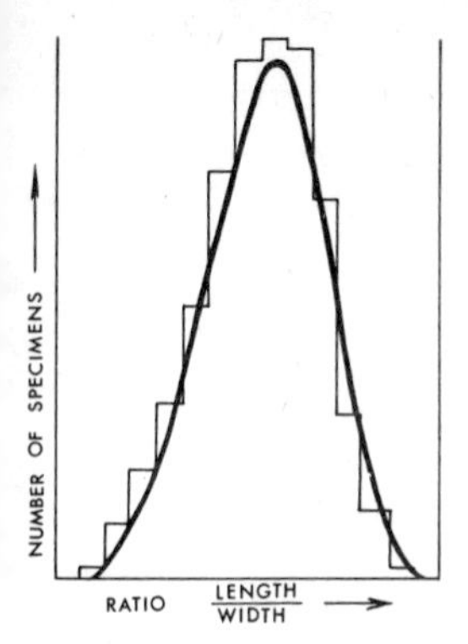

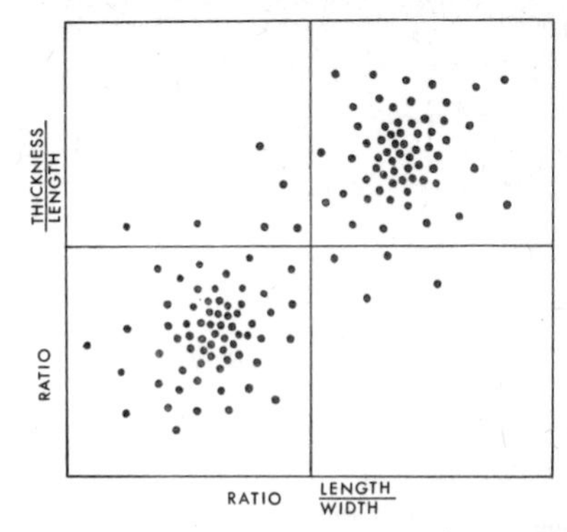

Fig. 40 *Two methods of analysing a fossil population. On the left – A histogram, the height of each polygon is determined by the number of specimens with a particular length/width ratio. Clearly only one species is present. On the the right – A scatter diagram. Each dot represents one specimen, the position of the dot being determined by the ratios of thickness/length and length/width of that particular specimen. The concentration of dots around two centres indicates that the members of this population should be referred to two species.*

members of human beings, who are all members of the species *Homo sapiens*. The permissible limits of variation for the members of a species of fossil organisms have usually to be determined on morphological characteristics alone. The interbreeding test can only be applied by inference. If one finds large numbers of apparently very similar fossils preserved on the same bedding plane or in a thin stratum, one can apply statistical methods by measuring for each fossil such features as length, breadth, thickness, etc. The ratio of say length to breadth are then calculated for each specimen and a variation diagram constructed (Fig. 40). Clearly the fossils should not be distorted by earth movements or by the compaction of the sediments. If the graph shows a normal variation curve with the one maximum, it is reasonable to assume that one is dealing with a fossil population of the one interbreeding species. But if the curve shows two maxima, then members of two species must be present. Statistical studies of this type have been made at a number of horizons, notably on colonies of brachiopods from the Carboniferous Limestone, on the 'mussel bands' of the Coal Measures (p. 151) and the gryphaeas (p. 149) of the Lower Lias and on the micrasters (p. 162) of the Upper Chalk, but it is exceptional to find suitable material, whilst the collecting of an adequate number of well-preserved specimens may well be an arduous task.

Much more commonly fossil species have to be defined entirely on morphological characteristics and then clearly personal experience and opinion enters into deciding whether or not two almost identical fossils should be referred to the same or different species. In other words, are the very slight differences between them of specific rank or are they but the variations that might occur between members of the same species? Similarly one has a comparable problem in deciding whether or not a number of species can be grouped together as members of the same genus and so on in delimiting the larger classificatory groups.

As palaeontological research proceeds and as new types of fossils are being found, so the limits of species and genera are constantly being revised. Frequently one worker will decide that too great a range of variation is being shown by the members of the genus he is investigating. He will wish to divide the old 'broad' genus into a number of more precisely defined new genera. For example the old genus of *Pecten* is now split into a number of genera, *Neithea*, *Entolium*, etc. Clearly this process has involved a change in nomenclature. The fossil formerly known as *Pecten quinquecostata* is now called *Neithea quinquecostata*. It is extremely difficult to keep up to date with all the changes resulting from the revision work of systematic palaeontologists. It is therefore customary to retain both the specific and the well known but now obsolete generic name and to write it in brackets, e.g. *Neithea* (*Pecten*) *quinquecostata*. One then knows that one is meeting an old friend under a new guise. But it is very confusing when both the generic and the specific name are changed, as for example when the purists insist that *Chonetes striatella* (No. 56), a well-known fossil of the Ludlow Beds, should now be known as *Protochonetes ludloviensis*. Often such changes, putting an excessive strain on one's memory, are due to over-rigid insistence on the Law of Priority.

LITERATURE

Unfortunately the work of the systematic palaeontologists is not published in any one journal, but is scattered through a great range of publications, partly because of the heavy cost of printing plates of the quality needed to illustrate the fine details of fossils. The monographs of the Palaeontographical Society deal with a number of fossil groups, e.g. the Ammonoidea of the Gault, British Corallian Lamellibranchia, the Dendroid Graptolites, the Ordovician Trilobites of the Shelve Inlier, etc. *Palaeontology*, the *Journal of Palaeontology*, *Palaeontologische Zeitschrift*, are some of the other journals which frequently publish papers on systematic palaeontology. To refer a fossil to its correct species, often requires specialist knowledge, access to a very well stocked library and considerable time. Also it may be necessary to develop the fossil further to reveal certain morphological features. There are a few very distinctive fossils, such as *Marsupites testudinarius* (Fig. 33) or *Douvilleiceras mammillatum* (No. 74) which can be recognized easily, but usually it is wiser for the non-specialist to attempt only generic identification. To quote the species, especially in the field, may sound impressive, but may very well be inaccurate. It is for this reason that only the genus has been given for the majority of the specimens illustrated in this book.

If one can visit a modern museum, such as those of the Institute of Geological Sciences, the British Museum (Natural History), the Manchester Museum or most University Geological Departments, one can often identify one's specimens by comparing them with the exhibits. Many of the smaller museums in county towns have geological specimens on display, but the names given may well be seriously out of date. Labels written in faded ink should be regarded with suspicion. It is usually easy to see from the type and layout of the exhibits, whether a museum is kept up to date or whether it is living on its past glories.

Certain of the books listed in the Bibliography are of value for identification. The three British Museum Guides, *Palaeozoic*, *Mesozoic* and *Caenozoic Fossils*, contain magnificent line drawings of 1162 of the commoner British fossils. They can be slipped into one's pocket and are a 'must'. Woods' *Palaeontology* contains detailed descriptions of a much greater number of genera, but only a few are illustrated. Morley Davies's *Introduction* and Swinnerton's *Outlines* are more selective, the former on morphology, the latter on evolutionary aspects. Both contain a number of line drawings of fossils, but neither gives the broad coverage of Woods. Moore's *Invertebrate Fossils* is a well illustrated American book and therefore includes many genera that are not found in Great Britain. Finally there is the Treatise on Invertebrate Palaeontology, the definitive work written by specialists of international standing. This is appearing in over twenty volumes, each costing several pounds. Each volume deals comprehensively with a particular fossil group, for instance the brachiopods comprise two volumes, with brief descriptions and good figures of all known genera.

If one finds a fossil that cannot be identified, it is advisable to show it to an expert at a museum or a University Geological Department. It is always possible that it may be a new discovery of considerable scientific importance. In the same way, valuable work can be done by anyone who systematically collects fossils in a particular area. Careful collecting, especially from temporary exposures, may provide data for a variety of geological purposes. The accuracy of a geological map is controlled by the data available to the surveyor. If fresh fossil evidence comes to light the map of the area may need changing in detail. A suite of fossils, carefully collected bed-by-bed, may be invaluable for studies of evolutionary series or, if related to the detailed lithology of the strata in which they occur, for palaeoecological work. There are many instances of amateur collectors finding fossils in beds until then regarded as unfossiliferous. The discovery of these fossils enabled the age of the beds in question to be fixed with precision.

There are many examples of amateur geologists, who commenced collecting fossils as a hobby. Then they became interested in a particular group or concentrated on a certain quarry or area, and became well known, in some cases internationally famous, through the study that they made of their discoveries.

Much valuable work can still be done in geology with a keen eye and skilful use of hammer and chisel. This is unlike so many other branches of science where the call is for more and more expensive equipment.

FOR FURTHER READING

Andrews, H. N. *Studies in Palaeobotany*, 1961. Wiley. 487 pp.

Ager, D. V. *Principles of Paleoecology*, 1963. McGraw-Hill. 352 pp.

Brinkmann, R. *Geologic Evolution of Europe*, 1960. Ferdinand enke Verlag, Stuttgart. 161 pp.

British Museum (Natural History) Handbooks.

British Palaeozoic Fossils, 3rd Edn., 1969. 208 pp.

British Mesozoic Fossils, 4th Edn., 1972. 207 pp.

British Caenozoic Fossils (Tertiary and Quaternary), 3rd Edn., 1968. 132 pp.

The Succession of Life through Geological Time by K. P. Oakley and H. M. Muir-Wood, 7th Edn., 1967. 94 pp.

History of the Primates: An Introduction to the Study of Fossil Man, by W. E. Le Gros Clark, 9th Edn., 1965. 127 pp.

Fossil Amphibians and Reptiles by W. E. Swinton, 5th Edn., 1967. 133 pp.

Fossil Birds by W. E. Swinton, 2nd Edn., 1965. 63 pp.

Dinosaurs by W. E. Swinton, 3rd Edn., 1967. 44 pp.

Colbert, E. H. *Evolution of the Vertebrates*, 2nd Edn., 1969. Wiley. 499 pp.

Colbert, E. H. *Men and Dinosaurs. The Search in Field and Laboratory*, 1969. Evans Brothers. 283 pp.

Davies, A. M. *An Introduction to Palaeontology*, 3rd Edn., revised by C. J. Stubblefield, 1961. Allen and Unwin. 322 pp.

Holmes, A. *Principles of Physical Geology*, 2nd Edn., 1965. Nelson. 1288 pp.

Kirkaldy, J. F. *General Principles of Geology*, 5th Edn., 1971 Hutchinson. 349 pp. Paperback.

Kirkaldy, J. F. *The Study of Fossils*, 1963. Hutchinson. 116 pp. Paperback.

Kummel, B. *History of the Earth*, 2nd Edn., 1970. Freeman.

Kummel, B. and Raup, D. *Handbook of Palaeontological Techniques*, 1965. Freeman. 852 pp.

Laporte, L. F. *Ancient Environments*, 1968. Prentice-Hall. 115 pp. Paperback.

McAlester, A. L. *The History of Life*, 1968. Prentice-Hall. 151 pp. Paperback.

Moore, R. C., Lalicker, C. G. and Fischer, A. G. *Invertebrate Fossils*, 1952. McGraw-Hill. 766 pp.

Pokorny, V. *Principles of Zoological Micropalaeontology*. Pergamon.

Vol. 1 (Foraminifera, Radiolaria etc.) 1962. 623 pp.

Vol. 2 (Scolecodonts, conodonts, ostracods etc.) 1965. 465 pp.

Swinnerton, H. H. *Outlines of Palaeontology*, 3rd Edn., 1947. Arnold. 393 pp.

Wells, A. K. and Kirkaldy, J. F. *Outline of Historical Geology*, 6th Edn., 1968. Allen and Unwin. 503 pp.

West, R. G. *Pleistocene Geology and Biology*, 1968. Longmans. 377 pp.

Woods, H. *Palaeontology-Invertebrate*, 8th Edn., 1946. C.U.P. 478 pp. Paperback.

INDEX OF GENERA

Reference numbers to coloured illustrations are in bold type; figure numbers of line illustrations are in italics; page numbers of text in Roman.

GENERAL INDEX

Dalí

THE LIFE AND WORKS OF

DALI

Nathaniel Harris

A Compilation of Works from the
BRIDGEMAN ART LIBRARY

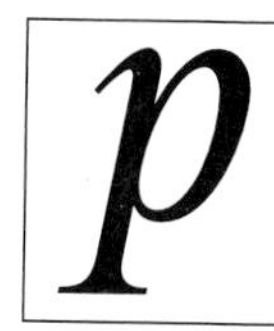

This is a Parragon Book
This edition published in 2003

Parragon, Queen Street House, 4 Queen Street, Bath BA1 1HE, UK

ISBN 1-85813-656-3

Printed in China

Editor: Alexa Stace
Designer: Robert Mathias

The publishers would like to thank Joanna Hartley at the Bridgeman Art Library for her invaluable help.

DALI 1904-1989

SALVADOR DALI WAS A GREAT ARTIST who was also a great self-publicist and showman. The combination was an irresistible formula for success. Dalí the showman, his moustaches arrogantly upturned, became a familiar figure to millions who had never been near an art gallery. In this guise he seemed always at the ready with a paean of slightly absurd self-praise or a string of wittily inconsequential remarks which might or might not be profundities. But those who scorned Dalí as a charlatan had to come to terms with the fact that he created a host of dazzling images, some of which, like the soft watches in *The Persistence of Memory* (pages 20-21), have entered the general consciousness of our culture.

Dalí was a Spaniard, born on 11 May 1904 in the little Catalan town of Figueras. In a sense, Dalí's entire world consisted of Figueras, the Ampurdán plain in which it stands, the little fishing village of Cadaqués just beyond the mountains, and nearby Port Lligat where he built his home. These are the settings for the great majority of his works, even when their foreground is occupied with a crucifixion or a civil war.

Though he acquired more than his share of childhood neuroses and sexual fixations, Dalí came from a solidly middle-class family. They had wealthy and cultivated friends who encouraged the young Dalí and kept him unusually well informed about the dramatic developments taking place in the world of art. He was already artistically well-equipped when he went to study painting in Madrid (1921-6), and the period was more important for the close friendships he formed with the poet Lorca and the director Luis Buñuel, with whom Dalí made the celebrated film *Un Chien Andalou* (1929).

From about 1927 Dalí was increasingly drawn to Surrealism. This Paris-based movement, influenced by the relatively new psychoanalytical theories of Sigmund Freud, created works dictated by the unconscious mind through dreams, automatic writing and other procedures aimed at freeing the artist from the tyranny of rationality.

In 1929 Dalí established himself as a member of the group with the help of Gala Eluard, the woman who became not only his lover and wife but his 'minder' and muse. Initially, Gala seems to have saved Dalí from a serious mental crisis, and without her support and belief in his genius he might never have achieved so much; on the other hand it was Gala, growing increasingly greedy and extravagant, who later encouraged him to commercialize and often trivialize his art. Dalí himself promoted an ever more extravagant cult of Gala, whose many appearances in his work culminated in almost goddess-like images.

Dalí painted his most famous, and probably his best, works in the decade 1929-39, using a 'paranoiac-critical method' of his own devising. It involved various forms of irrational association, notably using images which changed according to the viewer's perception of them, so that a group of fighting soldiers could suddenly be seen as a woman's face. A distinctive feature of Dalí's art was that, however bizarre the imagery, it was always painted in an impeccable 'academic' technique, with 'photographic' accuracy of a kind that most of his avant-garde artist contemporaries regarded as outmoded.

Towards the end of the 1930s Dalí was becoming known in the United States, where attitudes towards artistic innovation were less conservative than in the Old World. The outbreak of World War II and the German victory over France in 1940 prompted Dalí to flee to the United States, where he stayed for eight years. America provided abundant opportunities for Dalí to use his talents, and also brought out his exhibitionist side. He became a super-celebrity, staging 'happenings' long before the term was invented, and eventually even starring in TV commercials.

However, Dalí also continued to work hard and seriously, remaining prolific as an artist, designer and writer. He lived to become an icon of the hippie generation and to create a fantastic personal monument in the form of the Dalí Museum at Figueras, a total environment filled with bizarrely inventive objects and murals.

Dalí's later years were overshadowed by a degree of estrangement from Gala, although he was shattered by her death in 1982. Subsequently there was mounting concern about the number of fake works in circulation that were attributed to Dalí. He himself was partly to blame, since it was clear that he had been induced to sign hundreds, perhaps thousands, of sheets of blank paper which could obviously be put to illicit uses. Virtually a living ghost, he lingered on until his death on 29 January 1989. He is buried in the Dalí Museum in his native town.

◁ **Self-portrait with Raphaelesque Neck** 1921

Oil on canvas

THIS WAS PAINTED when Dalí was only seventeen, although he makes himself look somewhat older and uncharacteristically rugged. 1921 was the year when his mother died (according to Dalí, one of his most traumatic experiences) and when he left home for the first time to enrol as as student at the San Fernando Academy in Madrid. The rather severe, challenging self-portrait has a misleading air of machismo, probably intended to conceal Dalí's extreme timidity, as did his later and better known self-image as a mustachioed dandy and prankster. The painting technique, although accomplished, is still derivative, the brushwork and colour scheme showing the influence of Impressionism, Pointillism and other 'modern' movements which Dalí would soon reject in favour of a meticulously accurate 'academic' style. The background of the picture shows the sea, the Costa Brava coastline and the little fishing village of Cadaqués which were to figure so largely in Dalí's life and work.

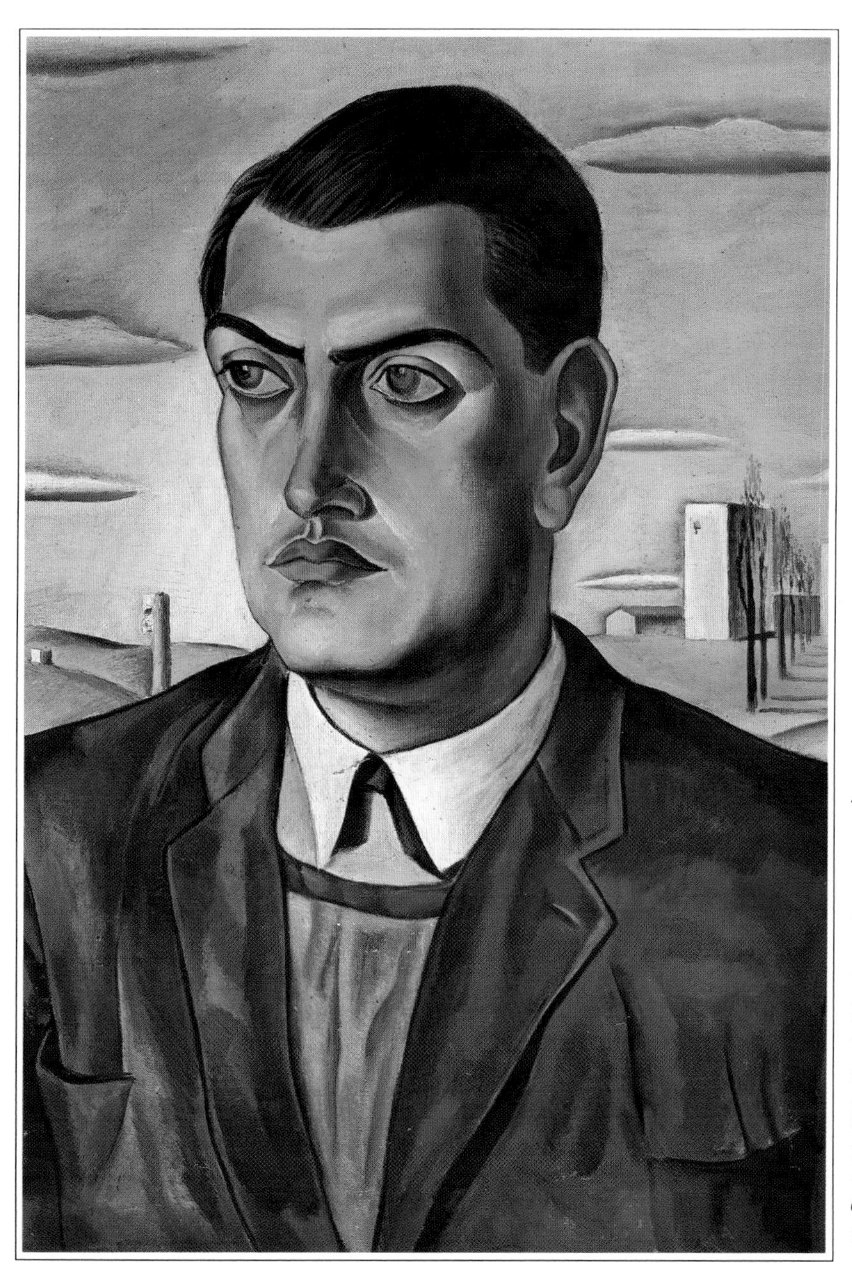

◁ **Portrait of Luis Buñuel** 1924

Oil on canvas

DURING HIS YEARS AS a student (1921-6), Dalí lodged in a large hostel in Madrid, the University Residence, where he formed close friendships with two fellow-students who would also become famous: Federico García Lorca, Spain's greatest 20th-century poet, and the distinguished film-maker Luis Buñuel. The rituals and eccentricities cultivated by the threesome had a permanent influence on Dalí's private mythology and public self-image. In the later 1920s, student cameraderie gave way to rivalries and jealousies, especially between Lorca and Buñuel. In 1929 Buñuel helped Dalí to establish himself in Paris and collaborated with him in making *Un Chien Andalou,* now the most celebrated of Surrealist films; the scene in which a girl's eye is abruptly sliced with a razor still has the power to make audiences gasp. Dalí's collaboration with Buñuel on a second film, *L'Age d'Or,* was less harmonious, and the friendship faded away.

▷ **Portrait of the Artist's Father** 1925

Oil on canvas

DON SALVADOR DALI, the notary of Figueras, was a man of strong personality who deeply influenced his painter son, mainly by evoking in the younger Dalí an intense and virtually lifelong reaction against everything he stood for. Don Salvador's commanding presence, and Dalí's helpless resentment, can be sensed in this and other portrait studies. In childhood Dalí rebelled through bedwetting, tantrums and bad performance at school; later, he deliberately failed to pass the final examination at his art college and gain the qualifications that would have assured him a 'respectable' future. After a complete estrangement between 1929 and 1934, Dalí and his father were reconciled, but the relationship remained an uneasy one.

◁ **Girl Standing at a Window** 1925

Oil on canvas

THE GIRL IN THIS PAINTING is Dalí's sister Ana Maria, who posed for several portraits in the summer of 1925. As in the equally well-known *Girl Seated* and *Woman at the Window in Figueras* (page 15), she is shown from behind so that her face is concealed; an oddity of Dalí's, particularly in evidence at this time, that has been given various psychological interpretations. The blanking out of the human element creates an effect curiously at variance with the holiday mood suggested by the glimpse of the river and the picture's notable lightness and clarity. The painting was shown as part of Dalí's first one-man exhibition at the Delmau Gallery in Barcelona, where it was seen and admired by Picasso. Ana Maria mothered the hopelessly impractical and sexually fearful Dalí, and was effectively his only female model until he met his wife-to-be Gala Eluard; Gala took over the role of model in an equally exclusive spirit, earning Ana Maria's undying enmity.

△ **Woman at the Window in Figueras** c 1926

Oil on canvas

IN THIS CHARMING PICTURE a woman sits making lace on a balcony overlooking the town square. Her equipment is shown in impeccable detail, and the scene can be interpreted as a Dalí counterpart to *The Lacemaker,* by the 17th-century Dutch artist Jan Vermeer. This painting became one of Dalí's obsessions, and visual references to it often crop up in his work. The treatment including the woman's modishly short, glossily highlighted hair is reminiscent of a magazine illustration in the reigning art deco style. In the background stand the buildings of Figueras, from first to last an important presence in Dalí's life: he was born and brought up there, his works were first exhibited at the town's municipal theatre, which he would later transform into a Dalí museum, and he was finally interred beneath its dome.

◁ **The Lugubrious Game** 1929

Oil and collage on canvas

BY THE LATE 1920S Dalí had embraced Surrealism and was developing the 'paranoiac-critical method' with which he plundered his psyche for images and associations. *The Lugubrious Game* (also known as *Dismal Sport*) was painted at Cadaqués in the summer of 1929, in preparation for Dalí's first one-man exhibition in Paris. Like several other works of the period, its subjects were masturbation, made overt by the statue's huge hand and Dalí's sexual fears and fixations. Among the objects in the spiralling column on the right is an image of Dalí himself, his mouth covered by a grasshopper, a creature of which he had an intense, irrational horror. Sexual and scatalogical images abound. *The Lugubrious Game* became the centrepiece of Dalí's extremely successful Parisian exhibition at the Goemans Gallery in November 1929. The painting was bought by his future patron, the Vicomte de Noailles, who hung it in his dining-room between works by Cranach and Watteau.

△ **The Great Masturbator** 1929

Oil on canvas

The title of this painting makes explicit the subject of *The Lugubrious Game* (page 14); Dalí himself described it as 'the expression of my heterosexual anxiety'. In 1929 he was still a virgin, inhibited by deep-seated fears of female sexuality and anal obsessions. Although these never left him, he seems for some time to have had a relatively normal sexual relationship with Gala Eluard, who left her husband, the poet Paul Eluard, for Dalí and eventually married him. *The Great Masturbator* was painted after their relationship had begun, but before they had unequivocally joined forces. According to Dalí, it was inspired by a 19th-century picture of a woman smelling an arum lily. Dalí moved the lily, replacing it with a well-endowed male figure. Woman, lily and male figure emerge from the strange 'Dalí' head which had already appeared in *The Lugubrious Game*. The small figure on his own, and the man making love to a rock shaped like a woman, point up the theme of solitary male fantasy.

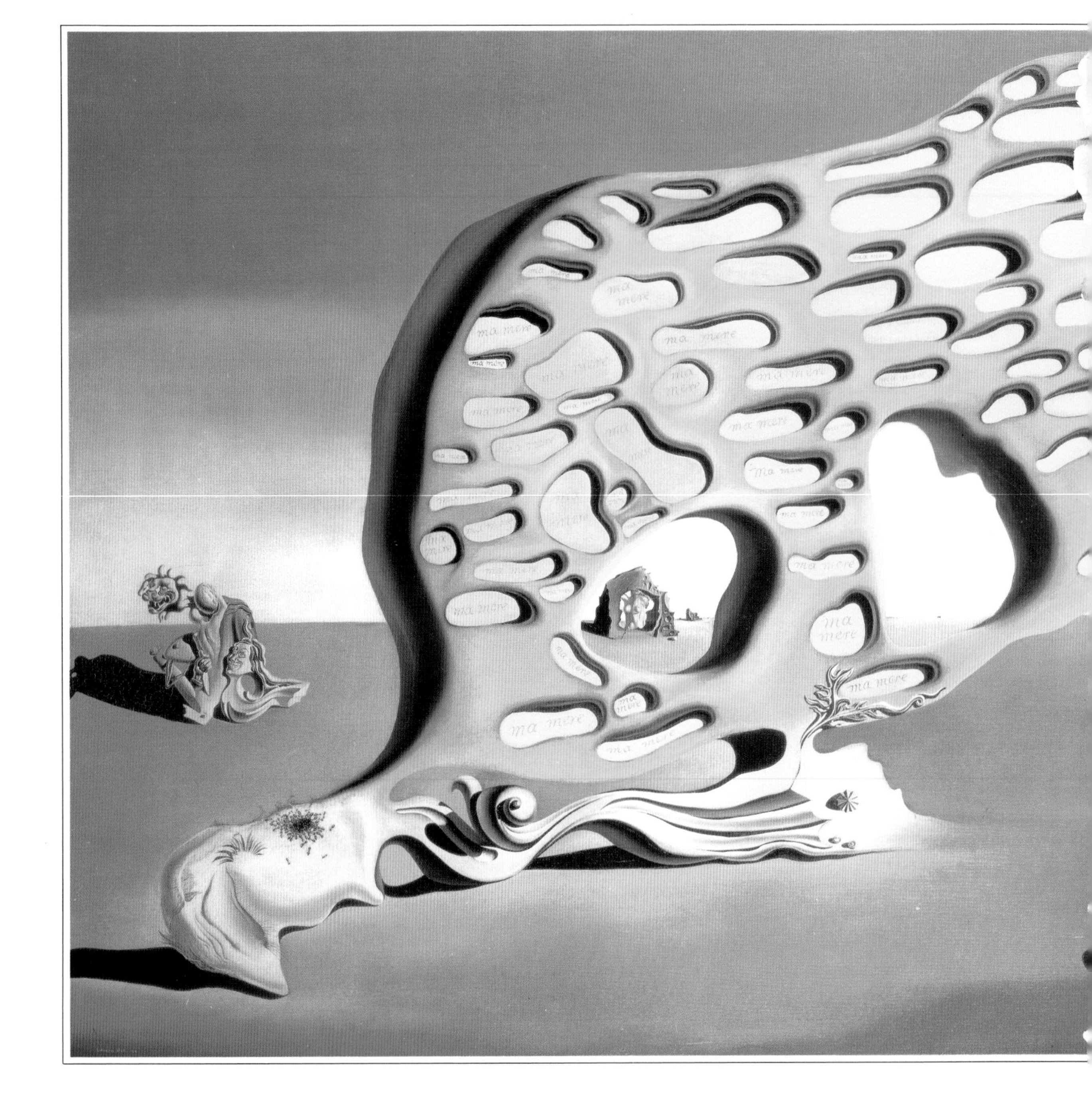
ma mère
ma mère
ma mère
ma mère
ma mère
ma mère
ma mère
ma mère
ma mère
ma mère
ma mère
ma mère
ma mère
ma mère

◁ The Enigma of Desire: My Mother, My Mother, My Mother 1929

Oil on canvas

LIKE *The Lugubrious Game* and *The Great Masturbator* (pages 14 and 15), this painting features a pallid, apparently soft head of Dalí himself, evidently asleep and presumably dreaming. It has been plausibly suggested that the shape of the head derived from the extraordinary rock formations around Cadaqués, where Dalí made his home. Here, however, the head and its familiar attachment of curlicued furniture are quite small, while the canvas is dominated by a large compartmented and pierced object reminiscent of a brain; many, but not all, of the 'cells' are labelled *ma mère* (my mother), doubtless because room had to be left for other obsessions! Ants, lion heads, a grasshopper, a sea creature with a shell, and other objects from Dalí's private world are present, along with two more obviously Freudian images: a hand holding a (castrating?) knife and a wounded female torso.

Detail

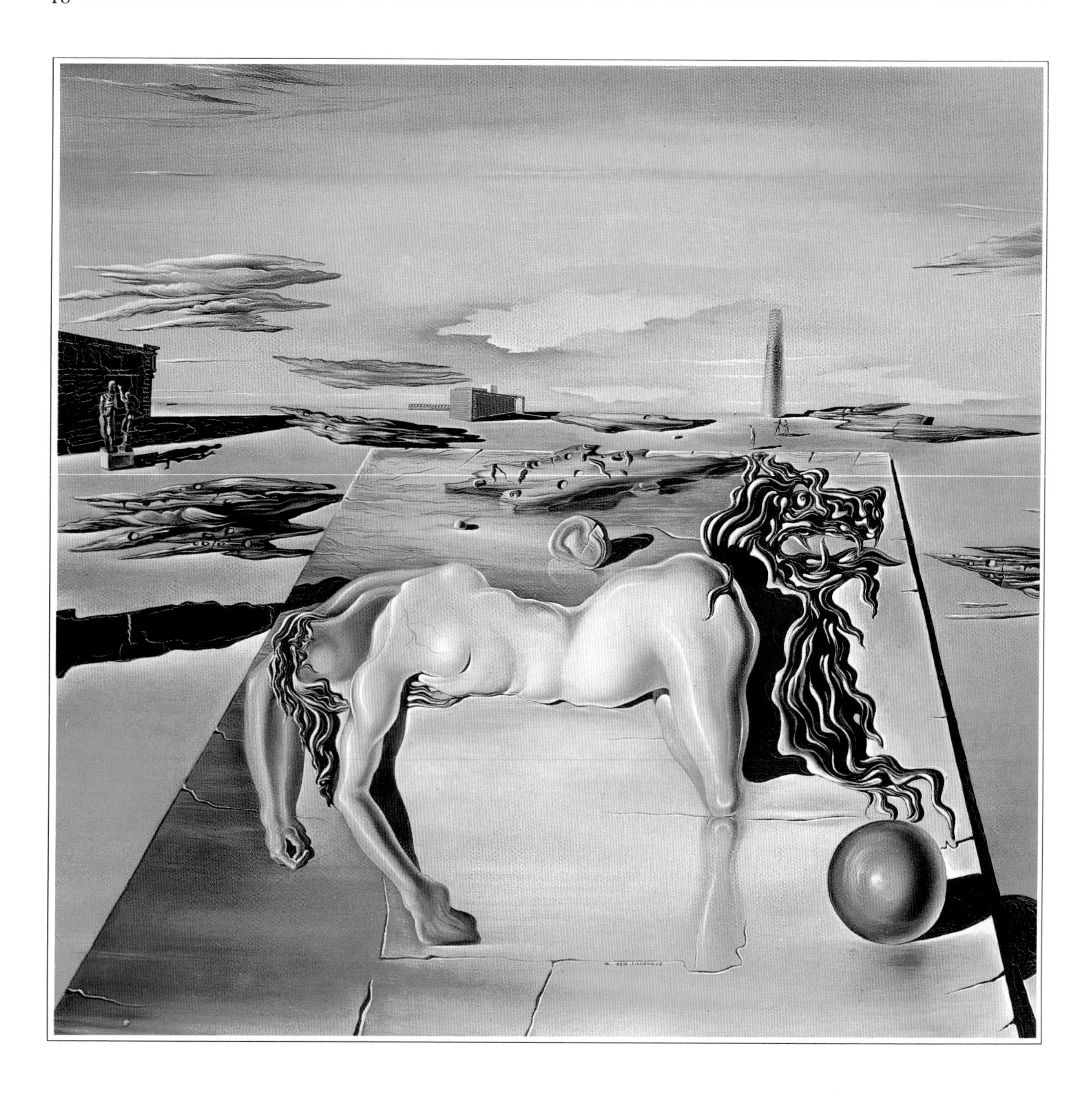

◁ **Invisible Sleeping Woman, Horse, Lion** 1930

Oil on canvas

DALI'S VERBAL DESCRIPTIONS of his 'paranoiac-critical method' leave most readers little wiser; but the personal effectiveness of the 'method' can hardly be questioned, since it enabled Dalí to tap his unconscious and produce chains of potent, startling images. By 1930 he was 'continuing the paranoiac advance' by devising a technique of registering multiple images which could be 'read' in different ways. This picture is an early example of a triple image: at its centre is a reclining female nude, but the maned head to her right transforms the image into that of a lion, while a shift of focus to the left brings out the image of a horse, its head formed by the woman's left hand and arm. This was a landmark in Dalí's development, signalled by the fact that he made no less than three versions of the picture: one was destroyed in a cinema foyer by right-wing demonstrators objecting to the Dalí-Buñuel film *L'Age d'Or.*

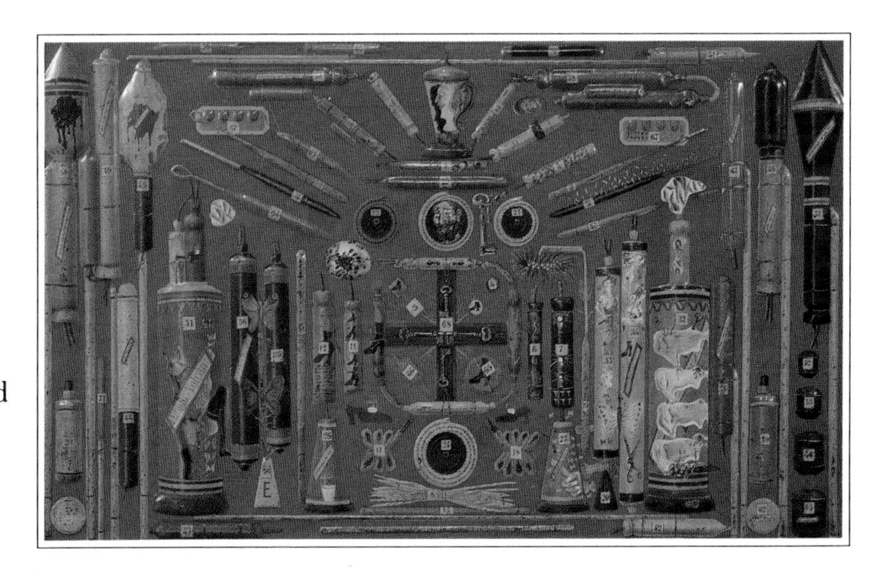

△ **Fireworks** 1930-1

Embossed and enamelled pewter

ALSO KNOWN AS *The Mad Associations Board, Fireworks* was originally a board advertizing the wares of a fireworks firm, amusingly transformed by the little paintings in oils which Dalí added to it. One of Surrealism's most audacious contributions to art was the 'found object' *(objet trouvé),* which might be a banal manufactured item (such as the bottle rack exhibited by Marcel Duchamp) or, more usually, a strikingly shaped natural object such as a rock. A further development was the 'assisted' found object, worked on by the artist; Dalí was characteristically inventive in creating Surrealist objects, which included not only *Fireworks* but a notorious *Lobster Telephone* and a sofa shaped like the red lips of the film star Mae West.

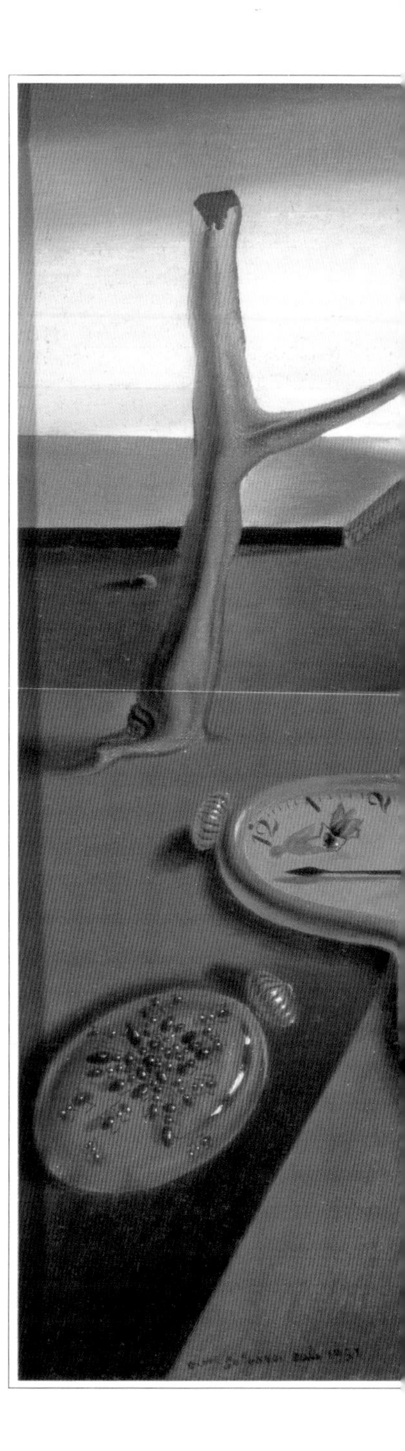

▷ **The Persistence of Memory** 1931

Oil on canvas

THIS QUITE SMALL PAINTING (24 x 33 cm/9½ x 13 in) is probably the most celebrated of all Dalí's works. The flaccidity of the hanging, slithering watches is a brilliant concept, more effective than many more sensational distortions in undermining our belief in a natural, rule-bound order of things. The imagery reaches into the unconscious, evoking the seemingly universal human preoccupation with time and memory. Dalí himself is present, in the form of the dormant head which had already appeared in *The Lugubrious Game* (page 14) and other paintings. Characteristically, he claimed that the idea for the painting came to him while he was meditating upon the nature of Camembert cheese; the Port Lligat background was already painted, so it took him only a couple of hours to finish the painting. When Gala, who had been out at the cinema, returned, she correctly predicted that no one who had seen *The Persistence of Memory* would ever forget it.

◁ **The Dream** 1931

Oil on canvas

THE DREAM WAS PAINTED at a time when Dalí and Gala were uncomfortably lodged in a fisherman's hut at Port Lligat which they were later to transform into a luxurious, labyrinthine home: not the least of the magical metamorphoses in which Dalí specialized. Paradoxically, in 1931 Dalí's pictures were having a considerable impact on the art world, while he was still finding it hard to make ends meet. At Port Lligat he worked untiringly, producing masterpieces such as this hauntingly atmospheric dreamscape. The image of a face without a mouth can be traced back to *Un Chien Andalou*, the classic Surrealist film made in 1929 by Dalí and Luis Buñuel; in it, the main male character literally wipes his lips away, in a gesture that was patently intended to be one of sexual menace. Dalí's eerie, claustrophobic dream world now looks familiar and 'natural' – perhaps because it belongs to the universal unconscious, or perhaps because images created by him have now been universally absorbed.

△ **The True Picture of the Isle of the Dead by Arnold Bocklin at the Hour of the Angelus** 1932

Oil on canvas

ARNOLD BOCKLIN (1827-1901) was a German Symbolist painter whose best known work, *The Isle of the Dead* (1880), was given its title by a dealer. Bocklin himself simply called it 'a picture to dream over', leaving the viewer to determine its meaning – an attitude close to the Surrealists' hearts. The relationship between the painting that inspired Dalí and his own 'Bocklin' canvas remains appropriately problematic. Something of Bocklin's romantic melancholy and taste for evening light lingers in Dalí's work, but he has replaced the lushness of the German artist's tree-crowded island with bare rocks, and the cup and the tall rod rising out of its bowl of course belong to the universe of Surrealism.

◁ **The Spectre of Sex Appeal** 1932

Oil on canvas

HUGE AND DECAYING on the beach, the rickety, propped-up figure of *Sex Appeal* is not a pretty sight; sex-appal might be a more appropriate term for this random collection of limbs, tenuously attached to a torso composed of sacks and ragged wrappings. The head merges into the rocks in a typical Dalí double image, and the amputated and ravaged extremities suggest the after-effects of cannibalism, an idea always closely linked with sex in Dalí's mind. According to his own account, the small boy in the sailor suit is Dalí at the age of six, holding a hoop in one hand and an ossified penis in the other. The identical figure reappears 35 years later in *The Hallucinogenic Toreador* (page 78). The cove is a real place, close to Cadaqués, which Dalí had probably known from his childhood.

▷ **Eggs on a Dish without the Dish** 1932

Oil on canvas

THERE IS NO ENGLISH equivalent to the punning French title of this picture: *Oeufs sur le plat sans le plat. Oeufs sur le plat,* which literally translates as 'eggs on the dish', is actually the French term for fried eggs; so 'eggs on the dish' can in fact exist without a dish, and even, as here, improbably hang from a line in mid-air, like bait on a fisherman's hook. Dalí was obviously pleased with the joke, since he repeated it in an identically titled canvas to which he added a further contradictory element by actually putting in a dish! His fondness for depicting two near-identical fried eggs, set close together and side by side, suggests that he was also aware of the slang use of *oeufs sur le plat* to describe small female breasts.

▷ **The Phantom Cart** 1933

Oils on panel

THIS IS A VERY SMALL PICTURE (19 x 24 cm/7½ x 9½ in), painted in oils on a panel. It is easy to overlook the fact that the central object is one of Dalí's multiple images, perhaps more successful than its equivalent in *Invisible Sleeping Woman, Horse, Lion* (page 18), since the illusion here requires no shift of focus to either side: the outlines of man and horse beneath the canopy of the cart can simply be reinterpreted as structures on the skyline of the town that the vehicle is approaching. So in *The Phantom Cart* the cart is actually less of a phantom than the creatures that draw and ride in it. The painting was based on Dalí's childhood memories of the day-long journeys between his home town in Figueras and the family house in the fishing village of Cadaqués, where he spent the summer holidays. For Dalí this was a period of freedom from constraint at home as well as at school, and this may help to account for the unusually serene atmosphere of the work.

◁ **Gala and the Angelus of Millet Immediately Preceding the Arrival of the Conical Anamorphoses** 1933

Panel

IN THIS ENTERTAINING psychodrama, a grinning, trendily dressed Gala confronts a distorted figure of the Bolshevik leader of the Russian Revolution, Lenin. Maxim Gorki, the famous Russian writer associated with the Bolsheviks, eavesdrops, oblivious to the lobster on his head. Dalí seems to have viewed Lenin as a father-figure, and as such an enemy to be ridiculed. Gala, whose long partnership with Dalí was now firmly established, appears in the role of triumphant liberator. The picture over the doorway is Jean-François Millet's *The Angelus,* a popular 19th-century image of peasant goodness and piety. It became one of Dalí's prime obsessions, into which he read erotic meanings (for example, that the man's hat concealed a state of urgent sexual arousal). Here *The Angelus* is shown more or less as Millet painted it; later it would undergo various metamorphoses at Dalí's hands (pages 29, 30, 31).

▷ The Architectonic Angelus of Millet 1933

Oil on canvas

ONE OF A SERIES of remarkable paintings in which Dalí pursued his obsession with Millet's apparently idyllic painting *The Angelus,* described in the caption on page 28. In a 1938 essay, 'The Tragic Myth of Millet's Angelus', Dalí described how the female figure in the picture became identified in his mind with sexual aggression, her attitude suggesting that of the praying mantis, which devours the male after copulating with him. The fantastic rock shapes on the coast at Cape Creus, close to Dalí's home, suggested a transformation of Millet's couple into enormous menhirs, or statues. Though the male (on the left) is larger, he is already under assault from the long 'needle' approaching his neck, perhaps a metamorphosed version of the pitchfork in Millet's original.

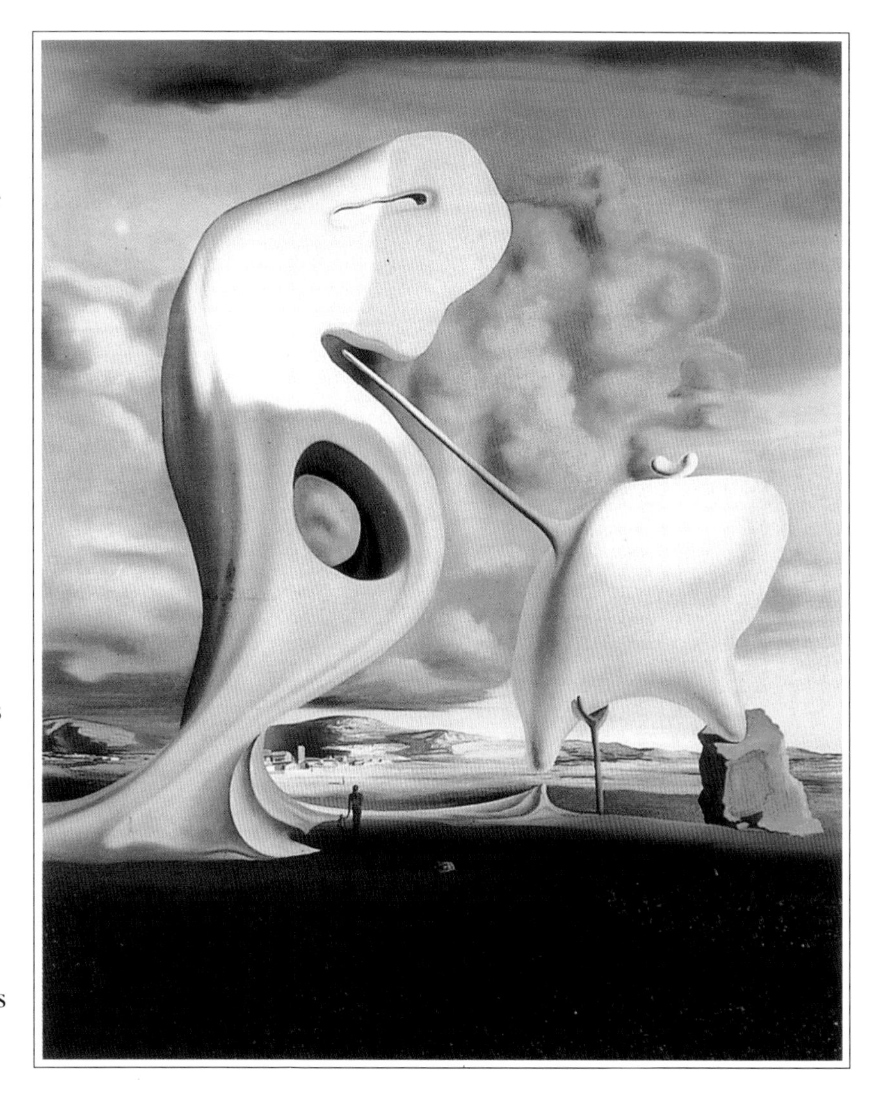

◁ **Atavistic Ruins after the Rain** 1934

Oil on canvas

THIS EVIDENTLY REPRESENTS a development of the scene in *The Architectonic Angelus of Millet* (page 29), in which the human figures have been metamorphosed into menhirs or sculptures. Actually they most closely resemble the sculptures of Dalí's contemporaries, Hans Arp and Henry Moore, but Dalí's treatment of them suggests that they are age-old rock formations. Dalí's own writings indicate that the 'remains' in this canvas are the result of erosion, but also of sexual conflict. The 'female' stone, which now dominates the picture, has devoured the male, in the process acquiring his characteristic void; all that remains of him is part of the display-like base of the object shown on page 29. The figures viewing the 'remains' represent the infant Dalí and his father, allied against the female sexuality of which Dalí had such a horror.

△ **Portrait of Gala** 1935

Panel

ALSO KNOWN, for obvious reasons, as *The Angelus of Gala:* behind her hangs a version of *The Angelus,* the painting by Millet that so obsessed Dalí (see pages 29 and 30); and the two people in the room – apparently front-Gala and back-Gala – are juxtaposed in roughly the same fashion as the peasants in the picture. But on this occasion Dalí has outrageously distorted Millet's composition, putting the peasants into a wheelbarrow (presumably as a form of erotic union) and altering the woman's attitude to increase the resemblance to a praying mantis that Dalí claimed to find in the original; presumably she is about to copulate with the man and then to devour him. By contrast, both Galas appear calm and untroubled, perhaps reflecting Dalí's conviction that she had been his saviour from sexual confusion and madness.

▷ **Mediumistic-Paranoiac Image** 1935

Oil on panel

BY THE TIME THIS WAS PAINTED, Dalí and Gala had formed a close relationship with the English collector Edward James, who contracted to buy the artist's entire output; the arrangement gave Dalí an assured income during the years when the Spanish Civil War (1936-9) had unsettled his life and he lived mainly in France. At one time James owned 40 important works by Dalí, including this one and a number of others illustrated in this book, for example *The Phantom Cart* (pages 26-27) and *Paranoiac-Critical Solitude* (pages 34-35). These were quite small (*Mediumistic-Paranoiac Image* is only 19 x 23 cm/7½ x 9 in), but Dalí also painted larger canvases, such as *Impressions of Africa* (pages 56-57), for James. Despite its title, *Mediumistic-Paranoiac Image* is almost Victorian in its realistic and innocently peaceful picture of the seaside, very much at odds with Dalí's normally subversive imagery.

◁ **Paranoiac-Critical Solitude** 1935

Oil on panel

HERE DALI TRANSPOSES real objects to conjure up impossible relationships as part of his self-proclaimed crusade 'to discredit reality'. Mysteriously abandoned, a car has become overgrown with flowers and plants. It appears to be integrated into the rock behind it, and the missing parts of the windscreen and body correspond with the large hole above it; yet the car is solid and separate enough to cast a shadow on the ground. To the left, the shape of the car is impressed into the rock face, and above it a plug of rock projects which corresponds to the hole on the right. It is as if a single rock has split and opened up, revealing a 'fossil' automobile which has been imprisoned in it for ages of geological time, an attractive explanation that does not, however, account for all the visual contradictions that Dalí has incorporated into the scene.

△ **Geological Justice** 1936

Oil on panel

LIKE *Mediumistic-Paranoiac Image* (pages 32-33), this is one of several small, cool paintings of empty beachscapes which Dalí painted for Edward James, who was supposed to be the artist's sole patron during this period. (In reality, Dalí's wife Gala sometimes cheated James by covertly selling paintings to other buyers.) The setting for these works was not Dalí's beloved Port Lligat, where he and Gala lived, but the nearby bay at Rosas, which had a very wide, flat beach. Although the scene looks perfectly natural and tranquil, the dark mud and stones form a figure resembling a steamrollered human figure, hence the original title, *Anthropomorphism Extra-Flat*. There is also a curiously sinister shadow on the right-hand edge of the scene, implying the presence of some possibly menacing figure. Nevertheless it is the atmospheric beauty of the work that remains uppermost in the viewer's mind.

▷ **Sun Table** 1936

Oil on panel

Like *Paranoiac-Critical Solitude* (pages 34-35), *Sun Table* shows Dalí moving away from purely erotic and autobiographical obsessions while continuing to set his surreal encounters against a local background, the empty Ampurdán plain and the coast around Cadaqués. The table is a humdrum object from a local café, supporting three coffee-glasses and a single coin. The tiles are a copy of those that were being installed in Dalí's Port Lligat house. These are mysteriously placed in a landscape with beached boats and a stretch of coastal sand that can evidently be reinterpreted as desert. Or at least that is what the presence of the camel suggests. Its egg-headed rider is faceless, unlike the bust which, though actually supported by a column, seems to rest on the camel's head. As on page 32, a piece of ancient pottery lies in the sand. Close to it, small but unmistakable, is another camel: on a pack of American cigarettes of a particularly well-known brand. The boy, seen in silhouette, may represent Dalí.

◁ **The Chemist of Ampurdán Looking for Absolutely Nothing** 1936

Oil on panel

THOUGH DALI'S IMAGE of him is spectre-like, the busy-about-nothing chemist was a real person, accurately depicted. Here he could well be a statue, like one of the mysterious monuments in the paintings of Paul Delvaux and Georges de Chirico, contemporaries who certainly influenced Dalí. The chemist makes another appearance in *Premonitions of Civil War* (page 42). The landscape is the Ampurdán plain as it appears in so many of Dalí's paintings, with its long perspectives and surrounding hills. As so often in his work, the title serves to undermine the content of the picture itself, making it irrational. Dalí pursued a similar strategy with regard to framing, insisting on putting pictures that were often quite small into ornate, heavy, old-fashioned frames which might initially deceive the spectator into believing that the subject would be treated in an equally traditional style.

Suburbs of a Paranoiac-Critical Town: Afternoon on the Outskirts of European History 1936

Oil on panel

▷ *Overleaf pages 40-41*

THIS IS A PAINTING of ingenious repeated images and neat dovetailing. In the foreground stands a cheerful Gala, holding up a bunch of grapes and welcoming the spectator into a scene that resembles three interlocked stage sets: all three were real places (the right-hand 'set' is the main street of Cadaqués). The grapes, the horse's skull and the statue's equine hindquarters are related shapes; but there are many more correspondences in the painting. The outline of the gateway behind Gala is a simplified version of the church tower that can be seen through it. The silhouette of the bell in the tower corresponds to that of the girl skipping in front of it. On the left-hand side of the panel, the low-domed structure, its high arcaded base and the two figures are reflected in the dressing-table, mirror and arch. Close by a miniature version sits on a soft open drawer.

◁ **Soft Construction with Boiled Beans: Premonition of Civil War** 1936

Oil on canvas

DALI LIKED TO CLAIM that this celebrated picture proved his intuitive genius, since it was finished six months before the outbreak of the Spanish Civil War in July 1936. However, Spain had been in turmoil for some years, and the origins of *Premonition of Civil War* can be traced to Dalí's experiences during the 1934 separatist rising in Catalonia. Though Dalí's work owes something to his great Spanish predecessor, Goya, he nevertheless succeeded here in creating a potent image of national agony: a towering, monstrous figure, its limbs jumbled and distorted, tearing itself apart. Dalíesque logic equated self-devouring with dining conventions, and so the ruptured flesh must be served up with vegetables, hence the scattering of boiled beans. Above one of the monster's hands we glimpse the Ampurdán chemist (page 39) still seeking absolutely nothing.

△ **Autumn Cannibalism** 1936

Oil on canvas

Autumn Cannibalism constitutes Dalí's response to the outbreak of the Spanish Civil War: a picture in which male and female carve up each other's flesh. However, the event is linked with the erotic obsessions so prominent in Dalí's earlier work, which reappear here in force. The cannibals are not fighting but kissing and embracing; the ants (a Dalí symbol of decay) are back; and the open drawer (another common Dalí image) implies the presence of the unconscious mind, a Pandora's box of unacceptable drives and impulses. Ironically, while Spain was being torn apart, Dalí was having a great success in the United States, dominating the Surrealist exhibition at New York's Museum of Modern Art and entering one version of the US Hall of Fame by appearing in December 1936 on the cover of Time magazine.

◁ **The Great Paranoiac** 1936

Oil on canvas

ONE OF DALI'S most stunning double images, *The Great Paranoiac* was painted after a discussion between Dalí and a fellow-artist, José Maria Sert, on the work of Giuseppe Arcimboldi, a 16th-century Milanese painter celebrated for portraits whose subjects were composed entirely of related objects (fruits, for example, or weapons). In similar fashion, but with more dynamic results, Dalí's smiling paranoiac dissolves into a turbulent scene in which men and women strike attitudes of grief or dismay. The double image is repeated, with variations, in the background to the left. To the right, by contrast, a group of exhausted figures seem to be trying to haul a boat across the sand, perhaps acting out one of the delusions that seethe within the brain of the Great Paranoiac.

Detail

▷ Swans Reflecting Elephants 1937

Oil on canvas

THE HALLUCINATORY IMAGES created by Dalí's 'paranoiac-critical method' are of two main kinds: single images that change according to mysterious laws of perception, like *The Great Paranoiac* (page 44); and groups of two or more images that are unlike as subjects but are revealed as having disturbing visual affinities, for example the bell and the skipping girl in *Suburbs of a Paranoiac-Critical Town* (pages 40-41). But in this canvas Dalí has combined the two in a virtuoso display of illusionism. The swans and tree stumps, reflected in the water, somehow take on the appearance of elephants; yet when the picture is turned upside-down the swans are transformed into elephants and vice versa! The soft, slippery surfaces and writhing forms (even the clouds seem organic) create a distinctly uncomfortable atmosphere, apparently at odds with the presence of the prosaic, palely loitering man.

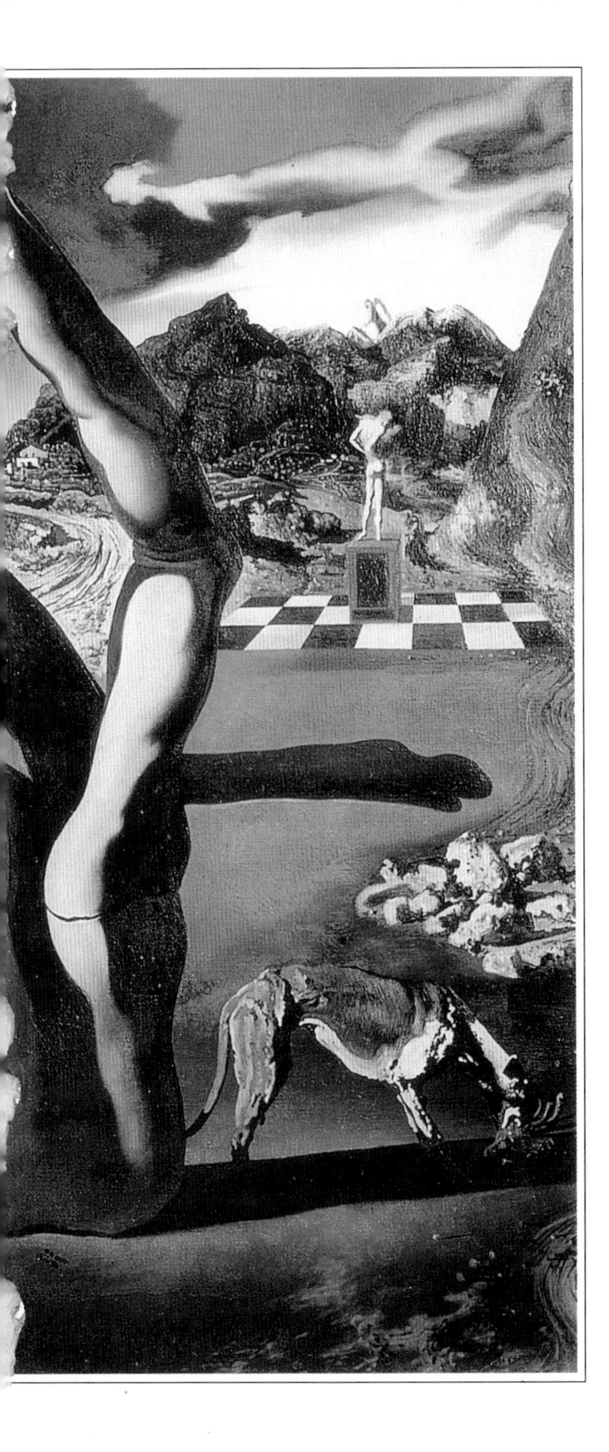

◁ **Metamorphosis of Narcissus** 1937

Oil on canvas

BACK IN PARIS after his great success in America, Dalí painted this picture and wrote a substantial poem to accompany it. In Greek myth, Narcissus was a surpassingly beautiful young man who saw his reflection in a fountain and fell in love with it. According to one version, unable to fulfil his desires, he pined away; but in a more dramatic alternative he leaned forward to embrace the image, toppled into the water and drowned. Afterwards the gods transformed him into the narcissus flower. Dalí shows Narcissus sitting in a pool, gazing down, while not far away there is a decaying stone figure which corresponds closely to him but is perceived quite differently – as a hand holding up a bulb or egg from which a narcissus is growing. In the background, a group of naked figures stand about attitudinizing, while a third narcissus-like figure appears on the horizon.

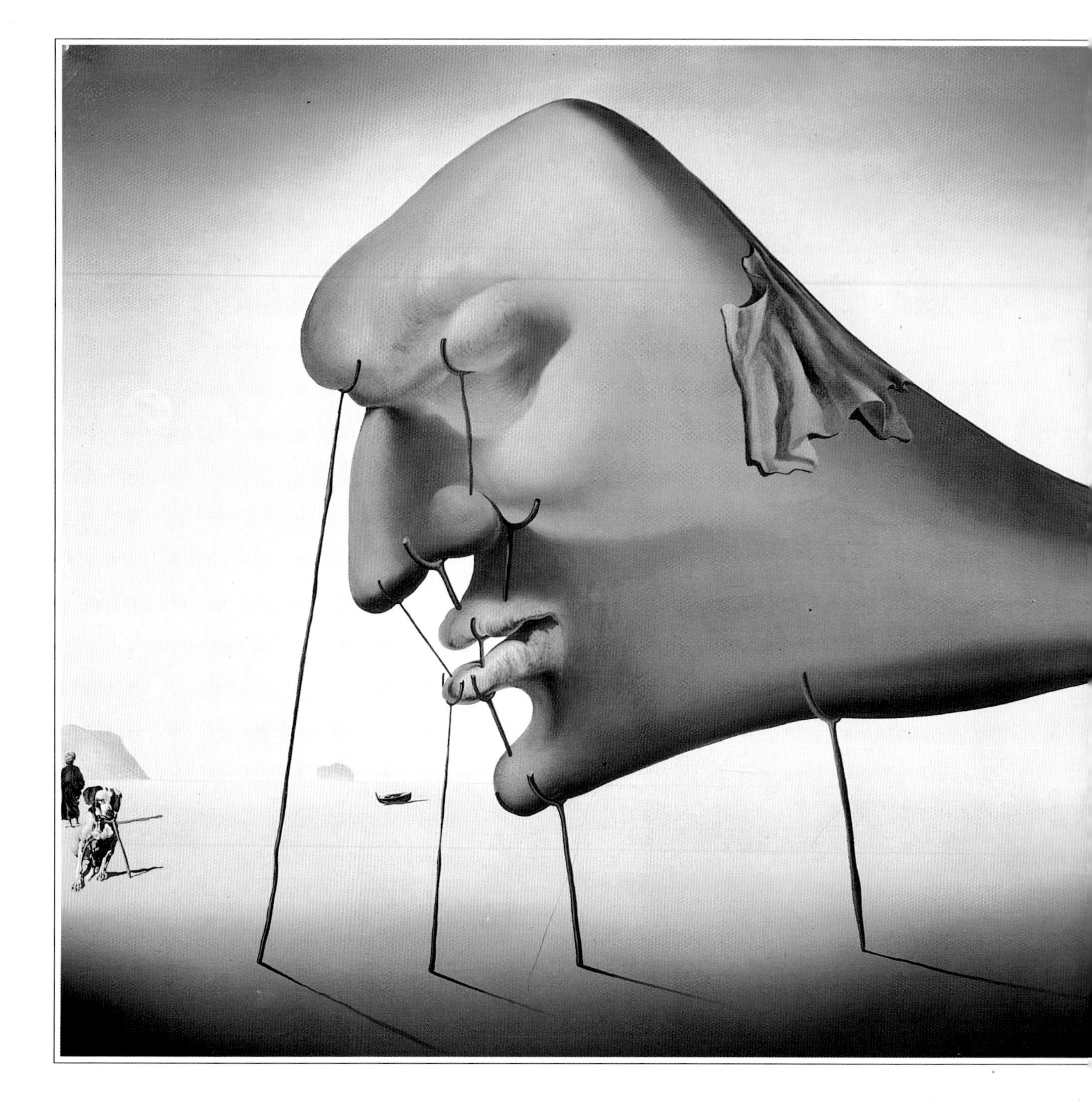

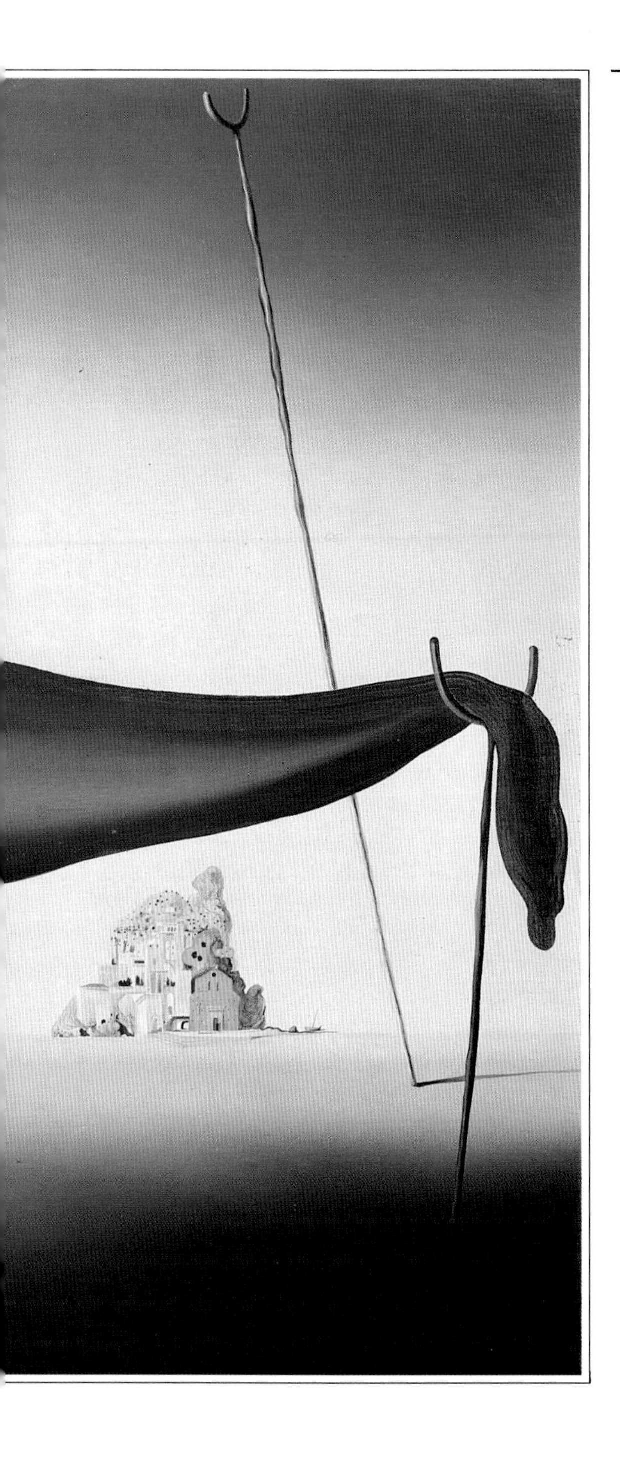

◁ **Sleep** 1937

Oil on canvas

IN *Sleep,* Dalí recreated the kind of large, soft head and virtually non-existent body that had featured so often in his paintings around 1929. In this case, however, the face is certainly not a self-portrait. Sleep and dreams are *par excellence* the realm of the unconscious, and consequently of special interest to psychoanalyists and Surrealists. Dalí's sleeper – or personification of sleep – is appropriately troubled, and an extraordinary number of crutches are needed to support the head and precisely position each feature. Crutches had always been a Dalí trademark, hinting at the fragility of the supports which maintain 'reality', but here nothing seems inherently stable, and even the dog needs to be propped up! Everything in the picture except the head is bathed in a pale bluish light, completing the sense of alienation from the world of daylight and rationality.

◁ **Palladio's Corridor of Thalia** 1937

Oil on canvas

DURING THE LATE 1930S Dalí spent long periods in Italy, where he made a careful study of Renaissance art. Among the masterpieces of the 16th-century architect Andrea Palladio was the Teatro Olimpico in Vicenza, in which both the theatre and the stage sets were modelled on ancient Roman examples. Here Dalí has devised an equivalent to Palladio's elaborate perspective effects by creating a receding corridor of human (if spectral) figures. The scene is divided into starkly contrasted dark and light areas, so that the corridor leads to the strongly lit figure of a girl playing with a skipping rope or hoop. This image had already appeared in *Suburb of a Paranoiac-Critical Town* (pages 40-41), and was based on a childhood memory of Dalí's cousin. Despite the innocent nature of her play, he somehow manages to make both the girl and her shadow as sinister as everything else in this curiously disturbing picture. Thalia was one of the nine Muses.

▷ **Spain** 1938

Oil on canvas

UNUSUALLY FOR DALI, the main image in this picture has a symbolic function which the artist himself draws attention to by painting in the title, *España* (Spain), at the bottom. Dalí's war-torn native land is represented by a woman whose head and upper torso can also be perceived as groups of fighting men; her lips correspond to the red cloak of one of the combatants, her nipples to the heads of two jousting horsemen. Both the woman's face and the fighters are painted in a style reminiscent of Leonardo da Vinci – appropriately enough, since Leonardo anticipated Dalí (and psychoanalysis) in recommending the study of moss, stains and cloud formations as stimuli from which the imagination could draw inspiration for new subjects. In his work, Dalí remained apolitical, dwelling on the self-devouring nature of the Spanish Civil War; in his life he was more opportunistic, becoming increasingly enthusiastic about General Franco as the Fascist forces advanced.

▷ **Mountain Lake** 1938

Oil on canvas

LIKE MANY OF HIS contemporaries, Dalí responded with alarm to the wars and threats of wars that plagued civilization in the late 1930s. According to Dalí himself, in *Mountain Lake* the telephone represents the apparatus used by Britain's prime minister, Neville Chamberlain, to negotiate with Hitler; at that date it was still a novel instrument of diplomacy. The fragility of the process is indicated by the presence of a crutch, and by the fact that the line has been cut. However, Dalí's political interpretation of the canvas smacks of hindsight, since in 1938 Chamberlain appeared to have achieved 'peace in our time' through the Munich agreement. The purely personal element in the painting remains strong, expressed through the striking double image of the lake, which can equally well be viewed as a fish on a slab or a male sexual organ!

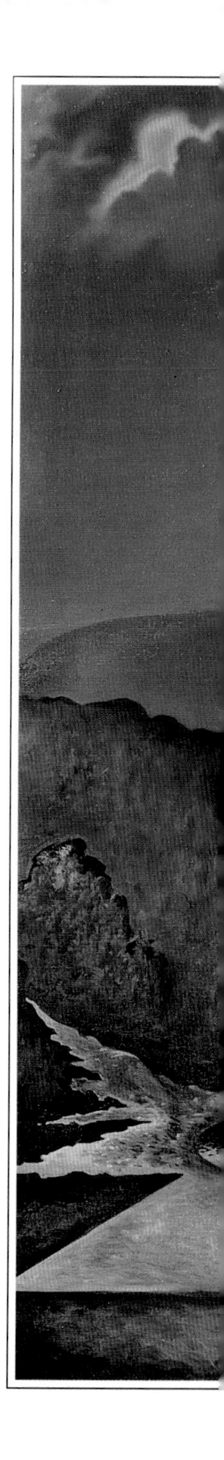

◁ **Impressions of Africa** 1938

Oil on canvas

THIS IS NOTABLE FOR the self-portrait of Dalí in front of his easel, staring fixedly in an effort to summon up images from his unconscious to transfer straight on to the canvas. His foreshortened hand, flung out at the spectator, is reminiscent of the 17th-century master Caravaggio, one of the Italian masters whom Dalí was diligently studying in the late 1930s. Typically Dalíesque double images are crowded into the back of the picture, including his wife Gala with eyes in shadow that can be interpreted as part of an arcade, and an image of a priest which also resembles a donkey's head. The African aspect of the work can be evaluated on the basis of Dalí's statement that 'Africa counts for something in my work, since without having been there I remember it so well!'

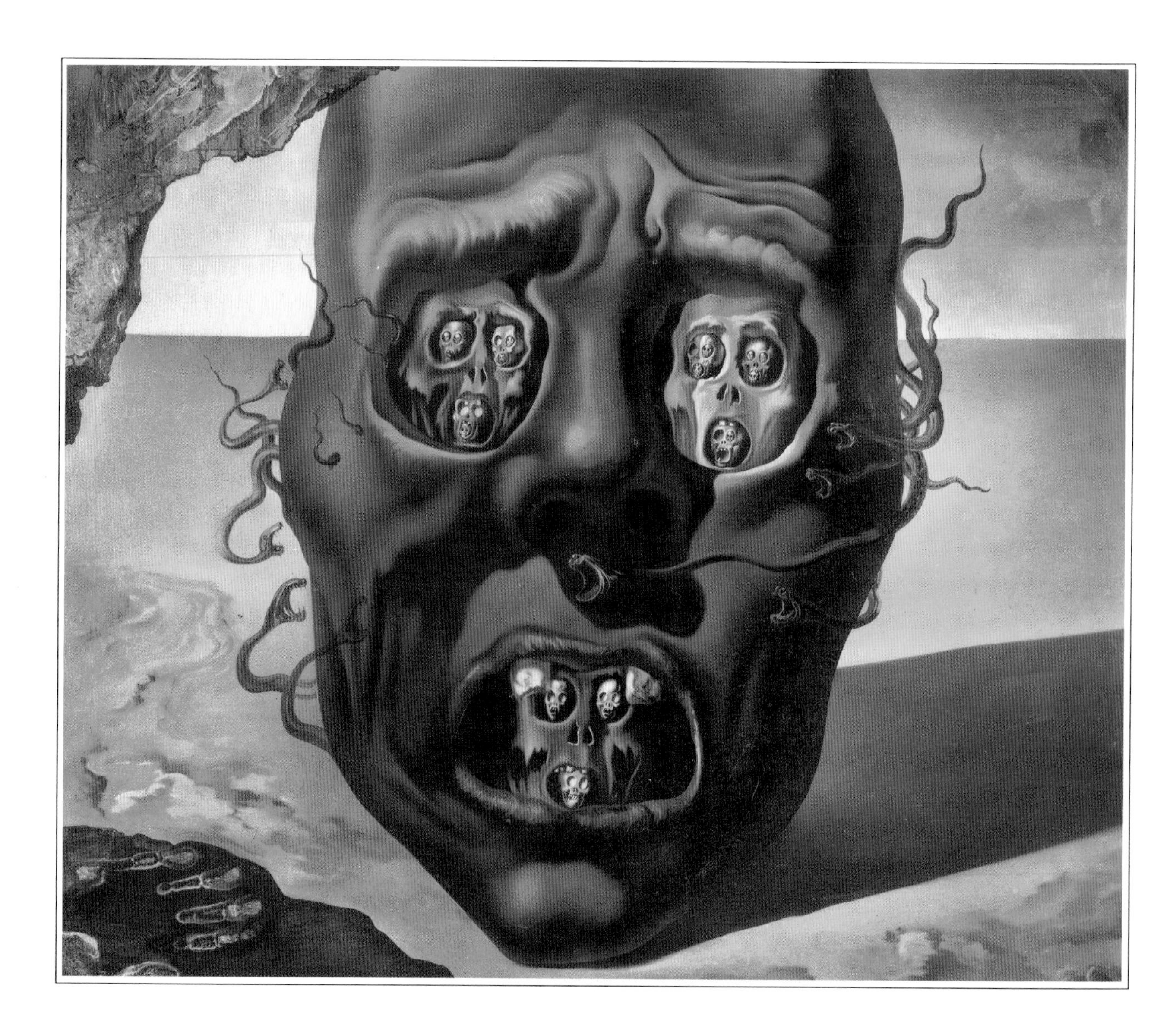

◁ **The Face of War** 1940

Oil on canvas

The Face of War was painted in the United States, where Dalí was to live for eight years and reach the pinnacle of his worldly fame and success. In the summer of 1940 he and his wife fled from France, whose armies were collapsing in the face of the German invasion, and, like so many refugees, sailed via Lisbon to the New World. The meaning of the painting is, for Dalí, unusually straightforward, employing symbolism rather than the irrational associations of the 'paranoiac-critical method'. A skull-like head surrounded by long, hissing snakes has every orifice filled with skeletons; each skeleton contains skeletons and skeletons-within-skeletons, so that the head is 'stuffed with infinite death', a potent symbol of the age of concentration camps and mass murders.

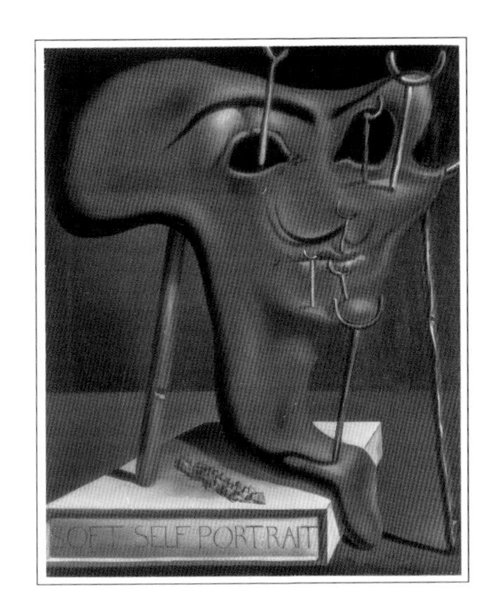

△ **Soft Self-Portrait with Grilled Bacon 1941**

Oil on canvas

USUALLY MOST IN EARNEST when dwelling on the subject of his own genius, Dalí has here caricatured his public image in a spirit of gentle self-mockery. His identity is mainly conveyed by the upturned, antenna-like moustaches which made his appearance instantly memorable. Crutches are an even more familiar presence in his work, and Dalí himself noted that the public, instead of growing tired of them, seemed to become more and more enthusiastic as he multiplied their numbers; so he has drawn the appropriate conclusion, using crutches large, medium and small to prop up his helplessly soft self-image on all sides. The ants round eye and mouth also signify decay or weakness. The grilled bacon lies in front of the self-portrait, which is made of a material which might well be excrement.

▷ Geopoliticus Child Watching the Birth of the New Man 1943

Oil on canvas

THE TITLE OF THIS PAINTING sounds like a parody of all the optimistic predictions made during World War II of the 'new world' that would emerge after the defeat of fascism. Geopolitics was a study, fashionable in the 1930s, which focused on the geographical factors influencing the destinies of states, and especially their location on the great continental land masses. The presence of a child watching the 'birth' of a grown man serves to undermine the concept and strengthen the viewer's scepticism, which Dalí shared. Like a chicken, the New Man is breaking out of the globe, which has a soft skin rather than a shell. Even the continents are soft and apparently on the point of oozing away; mysteriously, West Africa has shed a teardrop. The spiky canopy and the pointing woman, at once muscular and emaciated, help to give the picture its rather forbidding atmosphere.

Sentimental Colloquy 1944

Oil on canvas

▷ *Overleaf pages 62-63*

ON OCCASION, Dalí liked to claim that painting was the least significant aspect of his genius. It was certainly true that he assumed a great variety of roles, especially after his successes in the United States gave him abundant opportunities. He was, among other things, an inventor, a fashion and jewellery designer, a writer, a movie-set designer (Alfred Hitchcock's *Spellbound*), as well as a self-publicist, an early exponent of performance art and eventually the 'star' of many TV commercials. In 1944 Dalí wrote his only novel, *Hidden Faces,* worked with Hitchcock, and designed the sets for *Sentimental Colloquy,* a ballet loosely based on one of Verlaine's poems. Staged in New York, *Sentimental Colloquy* was a Surrealist extravaganza featuring dancers with underarm hair hanging down to the floor, a large mechanical tortoise encrusted with coloured lights, and the manic cyclists commemorated in this painting of the event.

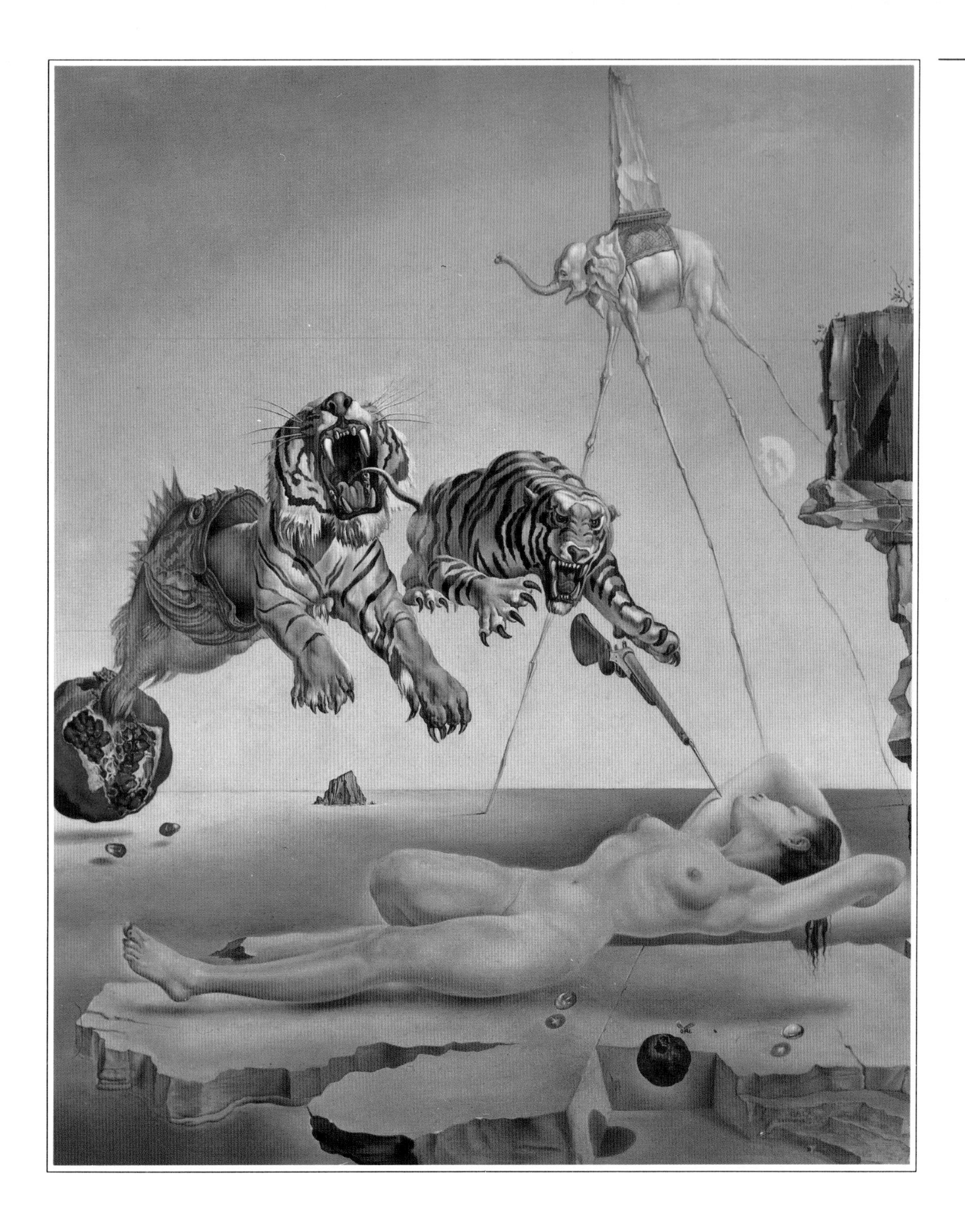

◁ **Dream Caused by the Flight of a Bee around a Pomegranate One Second before Waking Up** 1944

Oil on canvas

ONE SOURCE for this painting was a poster image of circus tigers, and Dalí has retained much of the bright immediacy associated with poster art. The bee and pomegranate referred to in the title are quite small objects, just below the outstretched body of the sleeping woman. She is obviously yet another portrait of Dalí's wife Gala, shown floating above (rather than resting on) a stone ledge, surrounded by the sea of the unconscious. The 'real' bee and pomegranate are dwarfed by the images they have brought into being: a huge pomegranate, the fish that has burst out of it, and the two tigers, depicted in all their snarling ferocity, which erupt from the fish's mouth. A more orthodox Freudian image, the rifle with fixed bayonet, and the fantastic elephant with stilt-legs, complete the one-second dream, which has apparently not had time to disturb the sleeper's tranquillity.

△ **The Basket of Bread** 1945

Oil on canvas

IN 1941 DALI announced that he intended 'to become classic', a development foreshadowed by his study of Italian Renaissance artists, and one which perhaps reflected a feeling that he had exhausted psychic autobiography. From this time he turned increasingly to the outside world as a source of inspiration, although he continued to interpret it in 'paranoiac-critical' fashion. Unlike most pioneers of modern art, Dalí had always produced highly finished works, and in the 1940s his style approximated still more closely to the 19th-century 'academic' ideal, widely regarded at that time as utterly outmoded. The pull of tradition, which became increasingly strong, can be seen in this superbly painted, apparently simple work, in which the bread and basket seem lit from within, floating in darkness. The subject had been tackled by Dalí 20 years earlier, attracting him partly because of its affinities with the work of admired Spanish masters such as Velázquez and Zurbarán. Here it also foreshadows Dalí's ventures into religious painting from the 1950s.

◁ **First Study for the Madonna of Port Lligat** 1949

Oil on canvas

DALI FINALLY RETURNED to Spain in 1948, although by this time he was an international celebrity and would continue to spend a good deal of his time abroad. At Port Lligat he set to work on a Madonna, completed in 1950, which marks the beginning of his religious phase. The picture illustrated here is actually a highly finished preliminary study, but arguably its quiet emotion is more effective than the grandiosity of the final version. The Madonna was blessed by Pope Pius XII, to whom Dalí was presented in 1949. Dalí had already made a cult of his wife Gala as a semi-divine being, so there was an unseemly element in his portrayal of her as the Madonna, even leaving out of account her own marked taste for young men and large sums of money that were not always honestly come by. The influence of Renaissance art on Dalí's work was now at its height. The fragmentation of images in the picture reflects the 'nuclear mysticism' that is also apparent in *Exploding Raphaelesque Head* (page 68) and other works.

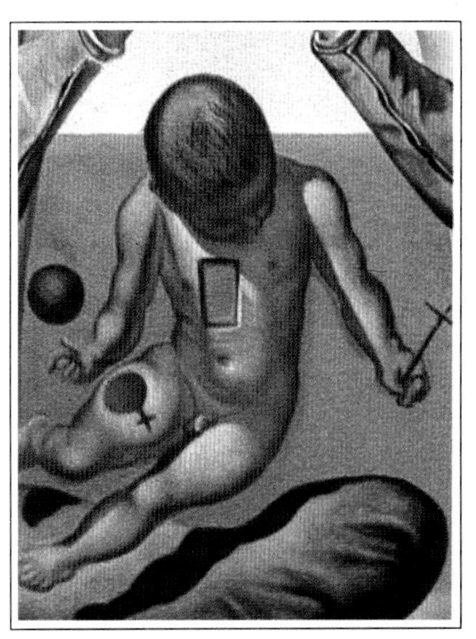

Detail

◁ **Exploding Raphaelesque Head** 1951

Oil on canvas

THE FRAGMENTED FORMS which appear in this painting and on page 66, originated in Dalí's study of nuclear physics. Deeply impressed by the discoveries which led to the development of the atomic bomb, he embraced 'nuclear painting' and 'nuclear mysticism'. Whatever their theoretical merits (or otherwise), these notions led to the creation of works of art in which Dalí's originality was very much to the fore. The head is like one of Raphael's Madonnas, classically pure and serene; at the same time it incorporates the interior of the dome of the Pantheon in Rome, with the light shining down through it. Both images are perfectly clear in spite of the explosion, which has blown the entire structure into small fragments shaped like rhinoceros horns, a new Dalí obsession most fully expressed on page 73. In the bottom left-hand corner, the fragments coalesce into a wheelbarrow, long an erotic symbol in Dalí's work.

▷ **Christ of St John of the Cross** 1951

Oil on canvas

A FAMOUS AND POPULAR painting, although there was a good deal of controversy when Glasgow Art Gallery decided to buy it in 1952. Its main source was a drawing made by the Spanish mystic St John of the Cross after a visionary experience. It showed the crucifixion, most unusually, from above, and according to Dalí this fused with his own 'cosmic dream' involving a sphere fitted into a triangle (Christ's head and the triangle formed by his arms and the line of the cross). The crucifixion takes place high over the rocky shores close to Dalí's home; but such an 'unhistorical' shift in time and place is common and acceptable in religious art, serving to emphasize the timeless, universal character of the New Testament narrative.

▷ **The Disintegration of the Persistence of Memory** 1952-4

Oil on canvas

DALI'S REWORKING of his famous *Persistence of Memory* (pages 20-21) is done in the spirit of the 'nuclear mysticism' displayed on pages 66 and 68. 'Persistence' and 'disintegration' might seem to be mutually exclusive, but not in the paranoiac-critical universe of Salvador Dalí. The soft watches are quietly falling apart, but much of the world about them seems to be fragmenting with production-line precision into geometric blocks; the straightened horns hint at the mathematical wonders of the rhinoceros (page 73). Most of the scene is under the water, which Dalí turns into a kind of skin, hanging from a branch; in other paintings it can be lifted like a sheet to reveal the sea-bed. Beneath the fish lies a transparent, near-extinct version of the self-portrait head that appears in so many works of the late 1920s and early 1930s.

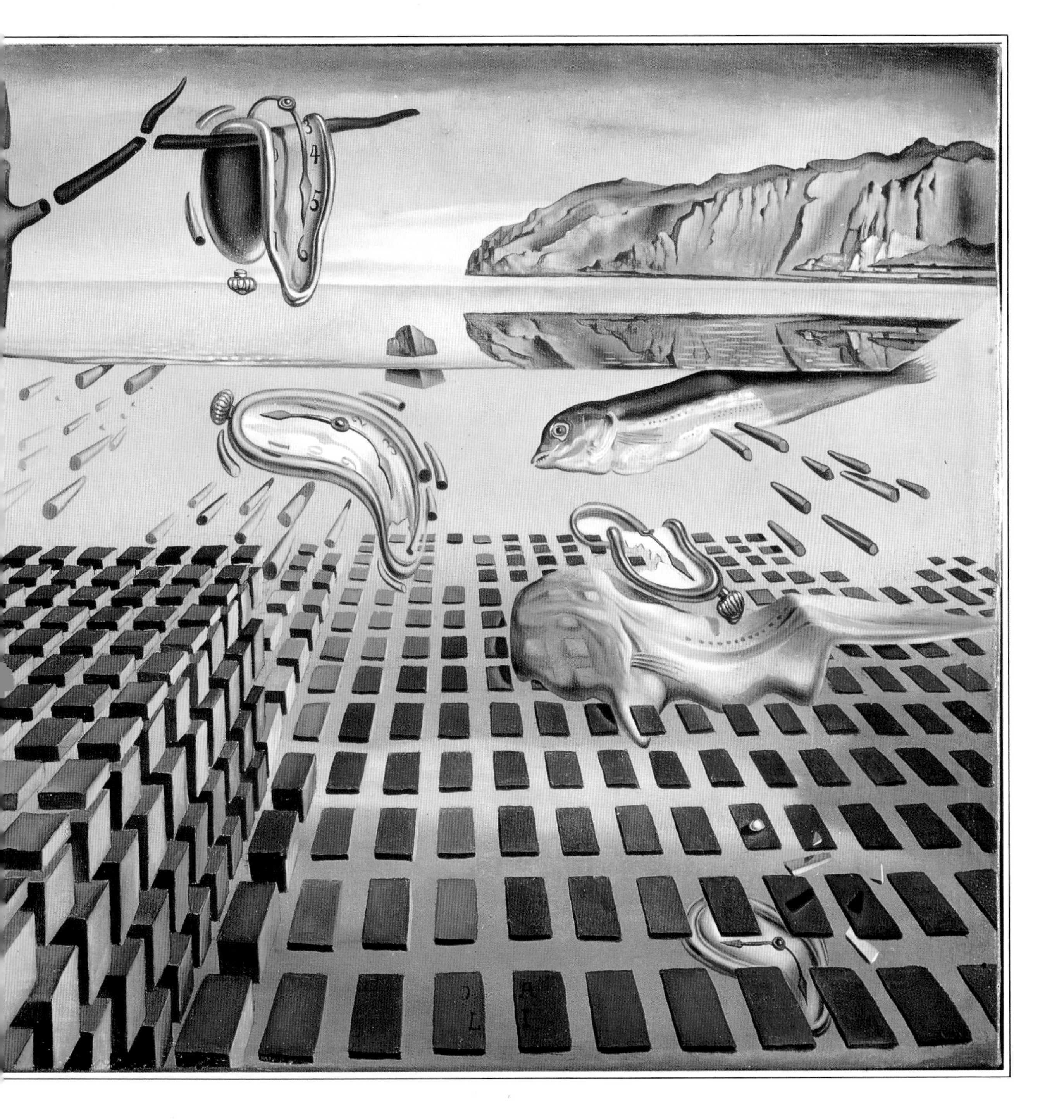

◁ **Rhinocerotic Figure of Phidias's Illisos** 1954

Oil on canvas

THIS COMES FROM what Dalí described as his 'almost divine and chaste rhinoceros-horn period', when he claimed that the curve of the beast's horn was the only perfect logarithmic spiral and consequently the ultimate in formal perfection. With characteristic Dalíesque logic – or critical paranoia – this insight came to him while he was copying a canvas that had obsessed him for decades: Vermeer's cool, lovely, light-filled portrait of a lacemaker.

In the mid-1950s Dalí even made a film called *The Prodigious Story of the Lacemaker and the Rhinoceros,* starring himself, a reproduction of the Vermeer, and a live, if carefully fenced-off, rhino. Here a torso from the Parthenon by the most famous of ancient Greek sculptors, Phidias, is fragmenting into a rhino head and horn-shapes which hang above a typical Dalí seascape, which is in turn suspended over the sea-bed.

◁ **The Last Supper** 1955

Oil on canvas

LIKE OTHER RELIGIOUS paintings by Dalí, *The Last Supper* provokes widely divergent reactions: some critics have denounced it as slick and banal, while others believe that Dalí has succeeded in revitalizing the traditional imagery of devotion. The controversies were complicated by public awareness of Dalí as a personality, apparently more interested in intellectual and emotional games-playing than expressing genuine convictions. Jesus and his 12 disciples are assembled within a modernistic, glass-encased room. The disciples, their heads bowed, kneel round a large stone table, their solid forms contrasting with the transparency of Christ. Two pieces of bread and a half-full glass of wine represent the sacramental meal. Dalí constructed this picture according to mathematical principles derived from his study of the Renaissance, and Leonardo da Vinci (who painted the most famous of Last Suppers) is a particularly strong influence. With a rather Leonardo-like gesture, Jesus points towards heaven and the figure (perhaps the Holy Ghost) whose arms stretch out to embrace the company.

▷ **Tuna Fishing** 1966-7

Oil on canvas

A VERY LARGE CANVAS, crammed with violent action, *Tuna Fishing* has an epic character that is unusual in Dalí's work. Its extraordinary photographic quality is a tribute to his skill, and also to his modernity in using a projector to place directly on to the canvas the images he wished to copy. The images themselves range from Hellenistic sculpture to the cinema. The slaughter of the fish is shown as a bloodbath which might equally well be a gladiatorial scene. The straining muscles and violent postures of the men are like figures by Michelangelo, heroically intense and glorying in the kill; by contrast, the authentic fishermen in the background are (literally) less colourful, going about their business with professional detachment. Also in the background, but by no means inconspicuous, a naked woman attracts unaccountably little attention.

◁ **The Hallucinogenic Toreador** 1968-70

Oil on canvas

THE ARTIST HIMSELF called this huge canvas, suggested by a box of Venus pencils, 'All Dalí in one painting', for it comprises an anthology of Daliesque images. At its top, a spiritualised head of Gala Dalí presides over the scene; in the bottom right-hand corner stands six-year-old Dalí, straight out of *The Spectre of Sex Appeal* (page 24). As well as many small images from earlier works, there is a Venus de Milo series in which the figure turns in the other direction and changes sex. The toreador himself is hard to make out at first, until we realize that the bare torso of the second Venus from the right can be interpreted as a face (her right breast as a nose, the shadow across her midriff as a mouth) and the green shadow on her drapery as a tie. To the left is the bullfighter's glittering 'suit of lights', which merges into rocks which can also be seen as the head of a dying bull. A metamorphical *tour de force!*

ACKNOWLEDGEMENTS

The Publisher would like to thank the following for their kind permission to reproduce the paintings in this book:

Bridgeman Art Library, London/Christie's, London 16-17, 32-33; **/Dali Museum, Beachwood, Ohio** 78; **/Davlyn Gallery, New York** 25; **/Ex-Edward James Foundation, Sussex** 26, 34-35, 36, 37, 38-39, 40-41, 44, 46-47, 48-49, 50-51, 52, 56-57; **/Index, Barcelona** 8-9; **/Index/Marquette University Fine Art Committee, Milwaukee** 66; **/Index/Museo Dali, Figueras** 65; **/Index/Private Collection** 15, 24, 59, 68, 70-71, 72-73; **/Juan Casanelles Collection**13, **Luis Bunuel Collection** 10; **/Museo de Arte Contemporaneo, Madrid** 11; **/Museo Espanol de Arte Contemporaneo, Madrid** 12; **/Museum Boymans-van Beuningen, Rotterdam** 53, 58; **/Museum of Modern Art, New York** 20-21, 31, **/National Gallery of Art, Washington D.C.** 74-75; **/National Gallery of Canada , Ottawa** 28; **/Paul Ricard Foundation, Bandol, France** 76-77; **/Perls Gallery, New York** 30; **/Perslys Galleries, New York** 29; **/Philadelphia Museum of Art, Pennsylvania, USA** 42; **/Private Collection** 14, 9, 22, 62-63; **/Reynold Morse Collection** 60-61; **/Tate Gallery, London**43, 54-55; **/Thyssen-Bornemisza Collection, Madrid** 64; **/Von der Heydt Musuem, Wuppertal** 23.